THE REMAKING OF RADIO

Broadcasting & Cable Series
Series Editor: Donald V. West, Editor/Senior Vice-President,
Broadcasting & Cable

Global Television:
How to Create Effective Television for the 1990s
Tony Verna

The Broadcast Century:
A Biography of American Broadcasting
Robert L. Hilliard and Michael C. Keith

Global Telecommunications:
The Technology, Administration and Policies
Raymond Akwule

Selling Radio Direct
Michael C. Keith

Electronic Media Ratings
Karen Buzzard

International Television Co-Production:
From Access to Success
Carla Brooks Johnston

Practical Radio Promotions
Ted E.F. Roberts

THE REMAKING OF RADIO

Vincent M. Ditingo

Focal Press
Boston • Oxford • Melbourne • Singapore • Toronto
Munich • New Delhi • Tokyo

This book is dedicated to the radio entrepreneur of the twenty-first century.

The author wishes to thank those executives who contributed their time, thoughts, and ideas to this project—especially Ralph Guild, chairman of The Interep Radio Store, and the Store's management and staff.

Focal Press is an imprint of Butterworth–Heinemann.
A member of the Reed Elsevier group

Recognizing the importance of preserving what has been written, Butterworth–Heinemann prints its books on acid-free paper whenever possible.

Library of Congress Cataloging-in-Publication Data
Ditingo, Vincent M.
The remaking of radio/Vincent M. Ditingo
p. cm.
Includes bibliographical references and index.
ISBN 0-240-80174-1 (acid-free paper)
1. Radio broadcasting — United States. I. Title.
HE8698.D58 1995 94-33921
384.54'0973—dc20 CIP

British Library Cataloguing-in-Publication Data
A catalogue record for this book is available from the British Library.

Butterworth–Heinemann
313 Washington Street
Newton, MA 02158-1626

10 9 8 7 6 5 4 3 2 1

Printed in the United States of America

The vacuum tube changed human society as fundamentally as those other great inventions we now take for granted—the wheel, the stirrup, the plow, the horse collar, the lathe, the number zero, the printing press . . . .

—*Arthur C. Clark*

Contents

Introduction xi
The Radio Timeline: A Historical Look at the Medium of Radio through the 1970s xv

PART I

1 **Prologue: Laying the Foundation for the Coming Years 1**

The 1980s: Age of Deliverance 1
The Target Medium 5
Enter the Satellite Age 6
The New Networks 10
FM Popularity 18
A Changing of the Guard 19
The Rise of the Mega-Reps and Mega-Agencies 22
Leveling the Field 26
A New Old Industry 27
Summary 28
Notes 29

2 **Radio's Fiscal Dynamics 31**

The Income and the Ratings 31
The Power Ratio 34

The Expenses 35
Average Salaries 36
Passing the Profit Threshold 38
Determining Value 39
A Three-Tiered Financing Approach to Radio Ownership 40
Tax Certificates and Minority Ownership 41
Notes 42

3 The Financial Boom and Radio's Emerging Entrepreneurs 45

Nouveau Radio 45
The Metromedia Deal 48
On the Bandwagon: Other Significant Transactions 48
The Turning Tide 51
What Went Wrong: Radio's Nadir 53
Notes 58

4 The AM Dilemma 59

Losing the Edge 59
Along the Road to Specialization 63
The Talk Explosion 65
Continuity 69
New Partnerships 69
Stereophonic AM 70
Regulatory Relief 72
AM Horizon 72
Notes 73

PART II

5 Realignment of the Industry 77

The 1990s: Unprofitability Yields New Blueprint for Ownership 77
Local Marketing Agreements 85
Impacting Personnel 87
Demographic Strategies 89

National Programming Shifts: The SyndiNets 91
Notes 95

6 **The Electronic Marketplace 97**

Managing Radio in the Information Age 97
The Local Radioscape 99
Notes 102

7 **The Digital Factor and Other Technological Applications 105**

Radio Broadcast Data Systems: Easing the Selection Process 105
Digital Audio Broadcasting 108
Direct (to Home) Broadcasting and Cable Audio 110
Digital Phone Transmissions 111
Notes 112

8 **Epilogue 113**

The Lessons 113
New Directions 115
At the Crossroads 117
The Next Chapter: Radio in the Twenty-first Century 117
Media Powerhouses 119
Establishing a Beachhead 120
One Final Thought 120
Notes 121

Selected Bibliography 123
Index 133

Introduction

This text is about a resilient form of communication that continues to entertain, inform, and stimulate the populace, despite having been all but written off by many national advertising agencies with the advent of television: It is about radio.

More precisely, this text is about the changing economic face of commercial radio broadcasting in the United States that began in the early 1980s with government deregulation of local station operators and the development of satellite technology for program distribution. These events have set the foundation for a new radio industry. The goal of this book is to provide a comprehensive examination of the many facets that shape modern-day radio, and to study a radio broadcast business environment that is constantly challenged.

In effect, what has occurred in the radio industry is a microcosm of late twentieth-century corporate America itself—a corporate America governed by consolidation—mergers, acquisitions—and specialization.

But this text is also a case study of adaptability. "Radio, more than any other medium in history, has enduring value because it has adapted to change," said National Association of Broadcasters president Eddie Fritts in a speech delivered during the association's annual radio convention in 1988. "Radio is the most consistent reflection of our country, in times of triumph and in times of trouble."

Indeed, today's radio industry is a tale of survival and tremendous growth underscored by a vociferous early period where many technological and patent battles were hard-fought among the medium's pioneering innovators, corporations, and organizations.

From my vantage point, I have seen the industry give rise to a new generation of leaders in the lender-friendly 1980s: Radio's *emerging entrepreneurs,* who, in many cases, successfully negotiated leveraged or management buyouts of established radio station groups. What separates these entrepreneurs from those of radio's early era is the fact that they were not solely driven by technology in order to compete.

From the free-spending eighties through the financial crunch of the early nineties, radio's most successful entrepreneurs have applied the disciplines of marketing, finance, and technology to owning and operating radio companies. They have reestablished America's oldest electronic communications medium as a major business enterprise within a burgeoning media marketplace.

Actually, radio's new owners can be considered the industry's third generation of entrepreneurs to advance the business, following both radio's pioneers and those post–World War II broadcasters who guided the medium in programming through television's soaring appeal.

These entrepreneurs have formed a new power structure in radio—one that also incorporates the remaining major radio broadcasting concerns, including ABC, CBS, and Westinghouse. They too have adopted innovative strategies to maintain their competitive stature.

This book explores the principal characteristics of radio's "new age":

- The initial deregulation of the industry by the FCC, particularly eliminating time-limit rules for ownership constraints, which resulted in a financial investment frenzy for radio;
- The introduction of satellite technology for the distribution of nationally produced programming from a central source to a local radio station, broadening the scope of network radio while spawning a trend back to national personality programs like those during the thirties and forties;
- AM radio's ratings decline with the explosive growth of new FM channel assignments in the seventies and eighties, along

with the attraction of FM stereo music programming, and AM's partial reprise in the nineties with the boom in talk programming;
- Efficient economies of scale for multiple station operations in the same market due to the FCC's 1992 decision to relax local and national radio station ownership constraints, thereby strengthening the financial stability of the industry in the face of languishing profits. This action also led to a major consolidation of ownership and staff positions.
- The integration of information technology for advertising transactions and radio's technological advancement toward digital audio transmissions designed to improve the medium's service to marketers and listeners, respectively.

The Remaking of Radio not only addresses these and other factors that have altered the course of the industry but presents views from several of radio's emerging entrepreneurs and experts on the forces guiding radio into the next millennium.

RADIO'S TIMELINE

The major business and programming changes that have been taking place in radio since the early 1980s were preceded by a wealth of history. In understanding the impact these changes have on creating a new radio broadcasting structure, it is important to highlight several key historical events that defined radio's role in both media and society. The *Radio Timeline* table that follows in the next section charts, in a quick-read design, the progression of the industry from its early beginnings through the 1970s. It is the preamble to *The Remaking of Radio.*

(Some of these past events have been critical to *The Remaking of Radio*, and their relationship is addressed in greater detail in the text.)

The Radio Timeline

A Historical Look at the Medium of Radio

BEYOND THE TELEGRAPH

The Early 1900s

1887 German scientist Heinrich Rudolf Hertz demonstrates that electromagnetic waves can be transmitted through space. Hertz's name was adopted as the measure of all radio frequencies.

1901 Italian engineer Guglielmo Marconi (who, in the late 1890s successfully demonstrated the use of a wireless telegraph [Morse Code] and formed the American Marconi Telegraphy Company) receives first trans-atlantic wireless signal sent from England to Newfoundland. The first practical applications of the new wireless are for ship-to-ship and ship-to-shore communication.

1904 British engineer Sir John A. Fleming develops the diode tube, which becomes the basis for transmitting wireless voice signals. It fails to amplify electronic signals, however.

1906 Inventor Lee de Forest devises a three-element or triode vacuum tube (known as the *audion*), which becomes the basis for radio signal amplification of voice transmissions.

Canadian inventor Reginald Fessenden transmits first distant voice and music broadcast from Brant Rock, Massachusetts, to ships at sea in the Atlantic.

1910 De Forest conducts experimental live radio broadcast from the Metropolitan Opera House in New York of *Cavalleria Rusticana* and *Pagliacci* with casts that featured opera tenor Enrico Caruso, marking the first significant wireless voice and music transmission in a U.S. city.

1913 American Society of Composers, Authors, and Publishers (ASCAP) forms to protect the creative works of musicians and music publishers airing over the new experimental radio stations. It is the United States' first music licensing organization.

1916 Inventor and business entrepreneur David Sarnoff, then commercial manager for the American Marconi Telegraphy Company, proposes developing a "radio music box" to receive radio transmissions.

1919 Radio Corporation of America (RCA) established primarily to market wireless radio receivers manufactured by General Electric and Westinghouse.

David Sarnoff eventually becomes the major force behind RCA's moves into radio broadcasting and television.

The 1920s

1920 Wireless receiving sets are advertised for sale to the public.

Westinghouse's KDKA Pittsburgh signs on by announcing the election returns of the Harding-Cox presidential race. KDKA is generally considered the first major U.S. radio station to offer regularly scheduled programming.

1921 Westinghouse's WBZ (Springfield, MA) receives first U.S. broadcast license granted by the Department of Commerce.

1922 First paid commercial "message" airs over WEAF New York. This station eventually became WNBC and then WFAN.

1923 First simulcast (a five-minute saxophone broadcast) by wire connects radio stations WEAF New York and WNAC Boston. The event marks the beginning of multiple "chain" station or network broadcasting.

National Association of Broadcasters (NAB) forms in Chicago to negotiate ASCAP's demands for music license fees from radio stations. NAB is now the broadcast industry's main trade organization and lobbying group in Washington, D.C.

First government classification of radio stations by signal power is undertaken to alleviate channel interference when broadcasting.

1925 Commentator Lowell Thomas makes first radio appearance over KDKA. Thomas is recognized as the pioneer of news broadcasting. He later joins the NBC Blue Network and eventually television.

1926 NBC (National Broadcasting Company) Radio Network broadcasts on twenty scattered stations through leased AT&T telephone lines, thereby becoming the first major radio network in the United States. Live music broadcast originated from New York's Waldorf-Astoria hotel. NBC Network, which was comprised of two different programming networks, NBC Red and NBC Blue, was cofounded by RCA, General Electric, and Westinghouse. GE and Westinghouse soon sold their interests in NBC to RCA.

1927 Congress establishes Federal Radio Commission (FRC) through the Radio Act of 1927 to regulate radio. One of its mandates was that stations operate in the "public interest, convenience or necessity." It officially designated the AM (amplitude modulation or standard wave) band for broadcast at 550–1500 kc, which has since been expanded to 1600 kc, and is going to 1705 kc.

1927–1928 CBS (Columbia Broadcasting System) Radio Network forms, first calling itself the Columbia Phonograph Broadcasting System, to compete head-on with NBC. Former cigar company executive, William S. Paley, becomes its president and eventually assumes the chairmanship.

1928 Advertising agencies expand beyond print ad placements to the radio market and gain a major foothold by handling network radio ads and supervising programs sponsored by their clients.

The 1930s

1930 Inventor Major Edwin H. Armstrong aggressively experiments with an alternative to AM band service—FM (frequency modulation) transmission, a system that would offer listeners greater fidelity with truer sound reproduction.

1932–1935 Comedians and musicians such as Fred Allen, Jack Benny, George Burns and Gracie Allen, Bob Hope, Bing Crosby, and Kate Smith rise to national prominence through weekly network radio programs.

1934 The Federal Communications Commission (FCC) forms as a direct result of the Communications Act of 1934 to regulate all communications, including telegraph, telephone, and radio. It replaces the FRC.

Mutual Broadcasting System is formed with four stations (WGN Chicago; WOR Newark, NJ; WLW Cincinnati; and WXYZ Detroit). It becomes the fourth radio network, competing with CBS and NBC Blue and Red.

1935 Major Armstrong demonstrates FM broadcasting with RCA, now led by Sarnoff. Armstrong's FM band uses a wider bandwidth than AM to achieve static-free signals with no interference from adjacent channels. Sarnoff, however, opts to protect AM's growth while turning his attention to the development of television, causing Armstrong to further the cause of FM radio on his own. Sarnoff's move delays FM's entry into the marketplace.

WNEW New York radio announcer Martin Block begins a show called "Make Believe Ballroom." Block borrows the title from a program hosted by KFWB Los Angeles announcer Al Jarvis, who hosted a similar show in 1932. The program intersperses records with commentary, signaling the birth of the radio "disc jockey."

1937 WLS Chicago reporter Herb Morrison records Hindenburg dirigible disaster in Lakehurst, New Jersey, on tape. It prompts the increased use of recordings and live on-site or "remote" coverage by networks.

1938 Orson Welles's fictitious CBS' *Mercury Theatre of the Air* "War of the Worlds" broadcast about Martians landing in New Jersey on October 30 causes many local residents to panic. The broadcast substantiates radio's influence over the American public.

1939 WQXR New York is key to early development of FM radio, providing the first regularly scheduled FM band programming through a relay to Armstrong's experimental W2XMN-FM station in Alpine, New Jersey.

General Electric begins marketing the first commercial FM radio receivers.

Broadcast Music Inc. (BMI) formed by radio broadcasters as an alternate music licensing organization to ASCAP in response to escalating ASCAP music rights fees. The not-for-profit organization's board of directors is composed of broadcasters. ASCAP and BMI have since emerged as the two leading music licensing organizations.

The 1940s

1940 FCC authorizes first FM radio stations.

1941 FCC issues its report on "Chain Broadcasting," aimed at protecting network affiliates from network control. Among other rulings, the FCC permitted affiliates of one network to use programs from other networks. The report also states that NBC's ownership of two radio networks—Red and Blue—constitutes a monopoly.

1943 NBC sells its Blue Network to businessman Edward J. Noble for $8 million. The sale eventually forms the basis for the ABC Radio Network in 1945.

1945 FCC moves FM band from its 42–50 MHz spot to its present 88–108 MHz frequency, allowing VHF-TV allocations to be divided into two portions of the same band—44–80 MHz and 174–216 MHz. The move makes all FM receivers and transmitting equipment obsolete, hindering the growth of FM.

News broadcasts during World War II lead to development of public affairs programming.

1948 CBS Radio lures away NBC's established talent lineup: Jack Benny, Burns & Allen, and Red Skelton—bolstering the fortunes of its radio and television networks. This becomes known in industry circles as "The Great Raid of '48."

The 1950s

1950 FCC, concerned about the increasing number of radio and television outlets in the United States, rules that no one entity can own more than seven AM, seven FM, and seven TV stations nationwide.

1950–1955 Network television adopts radio's popular variety and dramatic "block" format programming causing radio to become a mostly local medium offering music and news formats. According to FCC data, network radio revenues fell from $131.5 million to $64.1 million.

1954–1955 Formatted radio programming advances with the introduction of a new music genre that combines elements of country and rhythm & blues. Cleveland radio disc jockey Alan Freed calls the new music "rock 'n roll." The format is soon dubbed "Top 40" by radio programmer Todd Storz.

1955 FM broadcasting's evolution remains stunted due to a lack of listener interest coupled with a shortage of FM receivers. FM program offerings are generally classical music and noncommercial educational shows. Meanwhile, FCC permits FM stations to use their sidebands or Subsidiary Communications Authorizations (SCAs) to transmit what essentially is closed-circuit music.

1959 The National Association of FM Broadcasters (NAFMB) forms to promote the commercial viability of the FM band and the manufacturing of AM/FM home and automotive receivers. This group later becomes the National Radio Broadcasters Association (NRBA), merging with the NAB in 1984.

The 1960s

1961 Highly specialized programming is introduced by noted radio program innovator Gordon McLendon when he launches an all-news format over XETRA (AM) Tijuana, Mexico, which serves the Southern California region. The format would later be duplicated in many medium- to major-size U.S. cities.

FCC authorizes FM stereo broadcasting to foster the fledgling band's growth.

1962 FCC proposes new national allocation pattern for FM. Channel grants are made when an applicant makes a showing of need and technical feasibility. At the same time, FCC imposes a temporary freeze on new AM licenses.

FCC institutes three-year "anti-trafficking" ownership rule that stipulates that owners of radio and TV stations must hold onto their properties for a period of at least three years in order to ensure community service needs are met.

1964 FCC rules that FM stations co-owned in the same market as AM stations, providing the market has a population of more than 100,000, are limited to 50 percent duplication of AM programming. The simulcast restriction is designed for greater program diversity. (Simulcast restrictions have since been refined.)

1966 FM stations experiment with newer forms of rock music, called "progressive rock," offering listeners longer versions of popular music, album-cuts, and fewer commercials than AM "pop" music stations.

1968 ABC Radio launches four "demographic" networks out of its main network service to meet the diverse demands of the growing number of stations and formats. The FCC grants them an exemption from the long-standing network monopoly "chain" rule. This development in network radio became commonplace in the late 1970s and early 1980s.

1969 Contemporary rock station WABC-AM New York promotes itself as the most-listened-to station in the United States, underscoring "Top 40" music dominance on AM radio.

The 1970s

1970 The FCC, still struggling with media monopoly concerns, establishes its "one-to-a-market" rule prohibiting an operator of an AM, FM, or TV station in a given market from acquiring another broadcast property in the same market.

Automated tape/cartridge systems widely penetrate many FM studio operations, especially stations programming "beautiful music," a mostly instrumental format, or "easy listening" format, proving to be a highly economical investment for FM broadcasters. It leads to a number of new tape-supplied music services.

1975 The FCC institutes a media "cross-ownership" rule prohibiting future joint ownership of a daily newspaper and a radio or television station in the same market.

1979 Popular contemporary music formats migrate from AM band to both new and established FM stereo stations while news/talk and information formats find homes with AM audiences. Consequently, FM audience levels surpass AM levels.

REFERENCES

Sources for the *Radio Timeline:*

Broadcasting & Cable Yearbook, "A Brief History of Broadcasting and Cable," (R.R. Bowker, 1993).

Empire of The Air: The Men Who Made Radio, Tom Lewis (HarperCollins, 1991.)

The Broadcast Century: A Biography of American Broadcasting, Robert L. Hilliard, Michael C. Keith (Focal Press, 1992).

The Early Days of Radio Broadcasting, George H. Douglas (McFarland & Company, Inc. Publishers, 1987).

The First Fifty Years of Broadcasting (Broadcasting Publications, 1982).

The Radio Music Licensing Sourcebook (The Radio Music License Committee/Group W Radio, 1990).

Rebel in Radio: The Story of WQXR, Elliott M. Sanger (Hastings House, 1973.)

WNEW: Where the Melody Lingers On (Nightingale Gordon, 1984)

Part I

1

Prologue: Laying the Foundation for the Coming Years

THE 1980S: AGE OF DELIVERANCE

The 1980s left an indelible mark on radio broadcasting in the United States. It was, without question, a pivotal time in radio history as the industry underwent vast structural changes. It was also a time in which the aural medium experienced a renaissance of sorts.

Prior to 1980, many leading radio stations had been comfortably controlled by media conglomerates and large network broadcast companies that primarily were in the business of operating television properties. A major shift, however, occurred that decade in the media world. Television revenues, which are considerably higher than radio revenues, escalated further while cable and other new communications technologies experienced a growth explosion, leading many major broadcasting companies to divest themselves of radio and redirect their resources to these businesses.

No longer were the dominant players in the radio industry synonymous with traditional broadcast operations, such as ABC,

CBS, NBC, Metromedia, and Westinghouse. Mergers and acquisitions of networks and national representation firms along with management and stock buyouts of station groups created new power players. What emerged was a very different breed of radio broadcast owner—a *marketing and financial engineer.*

Many new owners also capitalized on modern technological applications, namely, the use of satellites for sending and receiving multiple program signals at the local station level.

It was radio's new entrepreneurial era, which rose to prominence during the lender-friendly 1980s and, in so doing, significantly influenced the medium's direction in business and in programming.

Entering the radio community were new corporate names including Apollo Radio, headed by former Radio Advertising Bureau president Bill Stakelin; Edens Broadcasting; Emmis Broadcasting; Evergreen Media; Infinity Broadcasting; Legacy Broadcasting; NewCity Communications; Noble Broadcast Group; Osborn Communications; Saga Communications; Spanish Broadcasting System; the Satellite Music Network; Unistar, and Westwood One, the latter three rewriting the rules of network radio.

The radio broadcasting industry, as many had known it, would never be the same.

A major factor contributing to this entrepreneurial environment was the deregulatory inclinations of the U.S. government; in particular, the Federal Communications Commission (FCC), the regulatory body for telecommunications in the United States since its founding in 1934. For radio, that meant the relaxation or elimination of many rules and policies governing ownership and nonentertainment programming, including the extension of the licensing period for radio stations from three to seven years and the suspension of public service requirements and commercial time guidelines.[1] Some radio broadcasters at the time felt the FCC action did not go far enough, saying it failed to address local and national ownership limitations and characterizing the ruling as "reregulation."

The initial deregulation order for radio took effect immediately after Ronald Reagan assumed office in January 1981. (Actually, the radio deregulation movement within the U.S. government began prior to Reagan's administration, under the reign of then-FCC chairman Charles Ferris and President Jimmy Carter in the late 1970s).

As part of deregulation, the FCC's so-called three-year "anti-trafficking" rule, which governed the buying and selling of radio stations, was eliminated. That rule, adopted in 1962, stipulated that with few exceptions, radio and television station owners had to hold onto their broadcast properties for at least three years to discourage quick for-profit ventures and foster local community service. In 1982 the FCC reasoned that the "broadcast service had matured to the point where the competitive environment could work to ensure that the station [facility] sales transaction could be regulated by marketplace forces . . . . A buyer who is willing to pay the market price for a station would be more likely to deliver the service audiences want than the owner unable or unwilling to continue station operation."

When this rule was eliminated, Wall Street financiers and the financial community in general began to view radio station licenses as marketable assets with potential investment value. Radio stations now had greater liquidity and became a more tradeable commodity. Investors also realized that radio, underpriced for many years, was a lower financial risk than other major businesses, particularly other forms of media.

Competition between radio stations to capture a larger portion of the local available audience—which leads to higher ratings and therefore higher advertising revenue—boiled over, as did competition between radio and local print media. Since there was no longer a time restriction on station trading (the exchange of ownership), the values of profitable, even marginally profitable, radio properties climbed rapidly. These events were further ignited by climbing interest rates and strong radio revenue growth. The status quo no longer applied and a financial revolution in radio broadcasting was underway.

Some of the more profitable radio stations in major cities in the United States that sold for $5 million to $7 million during the 1970s were sold for $40 million to $80 million by the late 1980s. Total dollar amounts for all commercial radio station sales in the 1980s alone, according to Paul Kagan Associates, a noted Carmel, California–based media research firm, climbed from $602 million in 1982 to $2.56 billion in 1985, when the FCC's initial deregulation moves began to noticeably take hold, to an all-time industry high of $3.45 billion in 1988. (See Chapter 5 for station trading cycle.)

The combination of easier access to financial backing and the decision by some long-established broadcasting groups to leave the radio business for other media and business opportunities opened the door for many radio broadcast acquisitions and buyouts.

Among those broadcasting concerns that left the radio station business during this period were NBC (from both network and station ownership), Doubleday, John Blair & Co., Katz Communications, LIN Broadcasting, Josephson International, Metromedia, and Harte-Hanks Communications. This resulted in a flurry of new station group buys, primarily by experienced radio managers who also found themselves among the industry's more active station traders.

Broader radio ownership was encouraged by yet another FCC deregulatory action. In 1985 the commission decided to expand U.S. national broadcast ownership restrictions by raising the existing seven AM, seven FM, and seven television station limit per entity to twelve each. This move ushered in a period of "group dominance," which, as discussed later in the text, was intensified during the 1990s when the FCC's concerns over ownership concentration lessened in the face of growing multimedia options.

It soon became evident that many of the new radio station deals consummated during the 1980s were too highly leveraged with considerable debt and, at the time they were transacted, contained overestimated financial market projections. Consequently, by the end of the decade some of the emerging new radio companies found it increasingly difficult to maintain certain loan covenants or obtain additional financing for other station or broadcast-related purchases. At the same time, credit ratings for many of these companies dropped. (See Chapter 3 for details on the financial impact of government banking regulations.)

This situation was further exacerbated as the United States fell into a major economic recession and credit backlash. The excesses of new buyers and lenders would soon steer the radio industry into an *entrepreneurial consolidation* phase.

What essentially occurred in radio ownership during the 1980s was a restructuring of the industry with the meteoric rise of new entrepreneurial radio owners and operators. They successfully demonstrated to investors that, with both radio's inherent fixed (operational) costs and already proven growth track record, the potential existed for a very appreciable business investment. What eventually transpired from all this activity was a remaking of the radio industry.

As a result of an expansionary radio economy, some longtime broadcasters questioned whether these *nouveau* radio groups could truly serve the public interest as airwave licensees. They raised the possibility that radio might find itself turning into a quick-profit business within a high-finance marketplace.

The subject set off a series of discussions in Congress that centered on whether there was just cause to reinstate the three-year rule or other ownership time constraints. But the FCC stood by its opposition to the old three-year ownership rule and in 1989 rejected a petition submitted by two citizen groups to bring the rule back. The following year, the U.S. Court of Appeals in Washington upheld the FCC's action.[2]

THE TARGET MEDIUM

One fundamental difference between modern-day radio and broadcasts before the mass introduction of television in the late 1940s is *program specialization*. In radio's early years, local stations relied heavily on nationally delivered network programming, referred to as *block programming*. Stations' daily programming agenda were made up of various nationally produced, popular shows that included live music, drama, and comedy along with several daytime serials and news reports. Television, however, became the primary purveyor of (national) block programming in the 1950s when its executives persuaded many then-popular radio personalities such as Jack Benny, Bob Hope, and Abbott and Costello to move to TV.

As the number of radio outlets continued to increase during the 1950s and 1960s, radio executives turned to specifically targeted or specialized local programming, known as formatic radio or *formats*, to effectively compete against television for audience share and against each other for advertising dollars. Radio networks redefined their role by offering mostly news and special event programming. Thus, radio broadcasting staked its claim as "the local medium."

One of the first successful local radio formats was Top 40 rock 'n roll programming, with the music carefully controlled by station executives. Top 40 formats helped radio, namely AM radio in the 1950s, recapture the attention of a visually influenced American public, particularly young adults lost to television.

Radio specialization became particularly evident during the 1980s with the creation of new, highly fragmented format designs ranging from all-sports to new wave jazz. Even Top 40 radio eventually fragmented into various programming forms to attract different groups of young adults. In some large markets, there can be as many as twenty-five individually defined formats, many of which are variations on major format categories. (See Figure 1.1).

Specialization is now occurring in magazines and in cable television. Due to radio's long-established targetability and immediacy factors for advertising placement, however, a discriminating advertising industry began to consider the medium as an efficacious marketing vehicle that could either complement other media or stand by itself. Radio again witnessed substantial advertising revenue gains.

ENTER THE SATELLITE AGE

Program specialization is only one component of radio's latest renaissance. As in most industries, technology has transformed the radio business. Indeed, there would not be a thriving radio broadcasting industry if it were not for technological innovations like the introduction in the early 1950s of the transistor, which allowed radio to become a portable medium.

During the 1970s, automated tape and cart machines allowed many local radio stations to operate more economically. By the 1980s, radio technology moved station operations to an even higher efficiency level with the use of satellites to transmit and receive nationally distributed programming signals. Satellite transmissions offered station operators a variety of quality programming with superior audio sound, particularly live and real-time events such as sports, concerts, and news.

In radio, satellites allow for the simultaneous distribution of multiple programming signals from a single source or national

CATEGORIES OF MAJOR RADIO FORMATS

Adult Contemporary/Soft Rock
Album-Oriented Rock/Classic Rock
All-News
All-Sports
Black/Rhythm & Blues
Classical
Country
Easy Listening
Ethnic
Jazz/New Age
MOR/Nostalgia
Oldies
Religious/Gospel
Spanish
Talk
Top 40/Contemporary Hit Radio
Urban Contemporary

Note: Within most major format categories are a number of niche formats designed for specially targeted audiences. Some formats such as all-sports and talk, and rhythm & blues and religious/gospel are closely allied.

FIGURE 1.1. Major categories of radio formats

origination point such as a network to a local subscribing station or affiliate. (A *program network,* in the strict radio broadcasting vernacular, serves a collective group of local stations—two or more—around the country or in a specific geographic region.) Satellite transmissions can be either analog or digital.

The major radio networks today transmit satellite signals to stations from "uplink" dishes or antennae—via the General Electric SATCOM Satellite System—in digital form because it allows for higher bandwidth efficiencies than analog-based systems. (See Figure 1.2.)

With digital satellite feeds, which transmit signals in bit pulses, the signal to noise ratio is vastly improved, resulting in superior audio reproduction quality and greatly enhanced signal

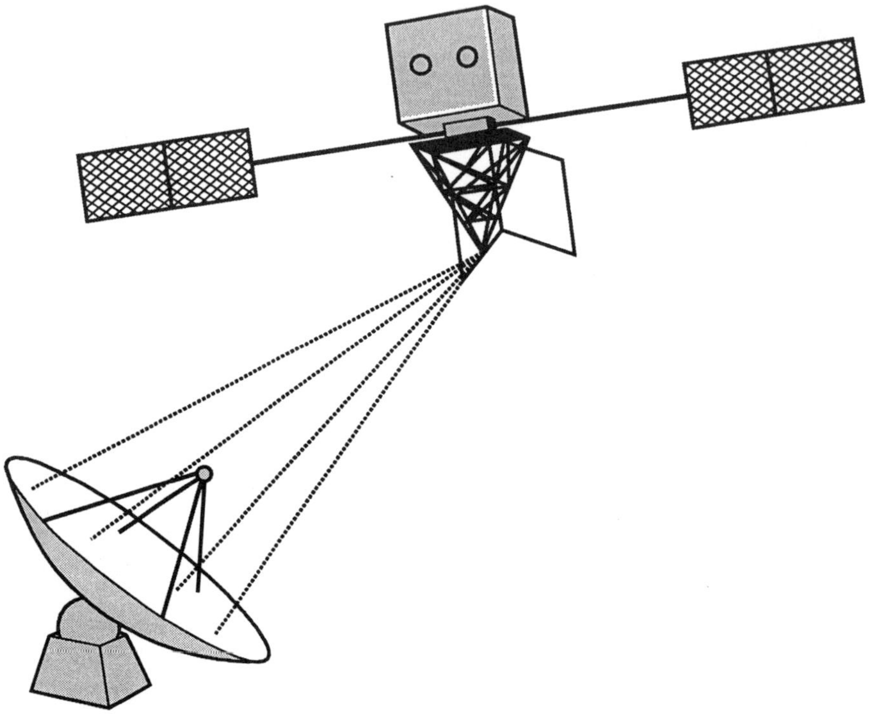

FIGURE 1.2. Network satellite model

reliability. Analog systems are more prone to interference because signals are transmitted in patterns analogous to the sound wave itself.

Before satellite transmissions, radio networks used telephone (land) lines for transmitting programming signals or feeds—one program at a time—to stations. Not only were satellite transmissions more flexible for network programming; they offered network executives and broadcasters favorable pricing advantages over leasing telephone lines.[3]

Beginning in 1979 and continuing through the early eighties, satellite technology became the U.S. radio industry's chosen method for interconnecting networks and program syndicators with their affiliated stations. National Public Radio, the upstart RKO Radio Network, and the Mutual Broadcasting System were among the first radio networks with an integrated satellite-to-affiliate hookup using dedicated uplink and downlink satellite antennae. Other national radio participants in the early stages of satellite-delivered programming included the Associated Press (AP) Radio News Network and the United Press International (UPI) Audio Network.

(Television networks began using satellites to transmit selected live events, mostly from overseas, during the early 1960s.)

These original satellite entrants used analog-based delivery systems, primarily through Western Union's WESTAR satellite, but eventually changed to digital when the big-three commercial radio broadcast networks—ABC, CBS, and NBC—moved to an integrated digital satellite system and corresponding downlinks in the early 1980s.

The application of satellite technology to radio programming resulted in the proliferation of live special event programming and nationally distributed music and talk shows. While one form of programming airs live, other satellite-delivered programming can be simultaneously recorded for playback later.

For both AM and FM stations, satellite technology advanced the *art of specialization* even further. It also marked radio's return to nationally produced programming for local consumption.

Initially it was no easy task, however, for radio networks to convince local radio executives that satellite technology was here to stay. Point-to-multipoint satellite-delivered radio programming was new and the concept appeared too futuristic to some station owners, who considered the technology more apropos to an episode from *Star Trek* than reality.

Resistance to satellite program transmissions permeated the industry as broadcasters rejected the added expense of buying and installing receive-only downlink antennae or "dishes," which averaged ten thousand to eleven thousand dollars each. (There was a moderate price reduction by the early 1990s). During the early satellite era some networks, including National Public Radio and Mutual, paid for their affiliates' dishes to boost satellite radio's growth.

While most radio broadcasters hesitated to jump on the satellite bandwagon, ABC, CBS, and NBC (who each required affiliates to pay for their own digital receiving dishes through an agreement with satellite equipment manufacturer Scientific Atlanta) were gradually switching their methods of program delivery from telephone lines to SATCOM. Under a "take it or leave it" policy, the major networks succeeded in replacing telephone land lines with digital satellite dishes as the conventional means by which radio stations received programming from national sources or networks.

The move toward a common digital satellite technology, commonly referred to as the Digital Audio Transmission System [DATS], set the standard for radio network programming, which remains in effect today. This standardization also allowed local radio stations to more easily have several simultaneous network affiliations.

Broadcasters soon realized that there was operational value to having numerous quality programming options at their disposal at any given time. From a purely economic standpoint, satellite technology gave station owners the means to save considerably on local programming and production costs, including local talent. (As explained in Chapter 2, savings will vary depending on the size of both the station and the market.)

THE NEW NETWORKS

In the traditional sense, network radio advertising means placing spots on a given network's lineup of station affiliates around the country, not accounting for individual ratings, and in many cases, format. This relationship between networks and affiliates usually entails a bartered time arrangement and/or compensation fees paid by networks for clearing commercial spots whether or not network programming is carried. To provide the greatest possible audience concentration for advertisers, the major networks in radio (as in television) own stations in the country's largest markets, allowing for easier program and commercial spot clearances in those cities. (In television, network compensation is also paid to station affiliates, but the carrying of regularly scheduled network programming is more common.)

Satellite technology soon led to a rapid expansion in the network radio industry to a new genre of entrepreneurial radio networking known as the 24-hour *format network*. By best estimates, some 25 to 30 percent of all commercially licensed U.S. radio stations in the early 1990s subscribed to a satellite format network.

Before satellite networks, most around-the-clock format programming had been on tape, delivered by mail to stations with automated tape-to-air equipment. Besides the means of distribution, the major distinction between these two programming categories is how they are sold. In general, the tape-distributed services contain no advertising and are usually sold on a cash-only basis. Most satellite services, on the other hand, are advertiser-supported, involving both program carrying fees and national advertising spot clearance time.

(A significant exception to tape distribution of around-the-clock programming before the satellite era occurred in 1975 when NBC launched a 24-hour news network, the National News and Information Service, also known as NIS, which was distributed to affiliates by land lines. The network, however, was discontinued nearly two years later due to both a general lack of local station support and NBC's failure to commit its own major market AM stations to the new service. Also complicating the situation was a decline in NBC's television ratings, which drained NBC's cash flow.)

The first satellite format network, Satellite Music Network (SMN) in Dallas, Texas, hit the airwaves in August 1981. Launched shortly thereafter was Transtar Radio Networks of Colorado Springs. Radio and television broadcaster John Tyler headed the Satellite Music Network effort; former Miller Brewing Company marketing executive C.T. Robinson started Transtar.

The idea behind these two new satellite programming ventures was to offer radio stations, principally in smaller markets, a variety of full-time quality music formats replete with announcers, jingles, and marketing support for a monthly fee and allotment of a certain bartered amount of minutes per hour for network commercial spots. There would be no on-air mention that the formats were national in origin. In addition, the subscriber station could take all or part (such as the nighttime hours between 7:00 P.M. and 6:00 A.M.) of the available 24-hour programming.

Similar to practices established by traditional radio news networks, these new satellite network services would sometimes pay select medium and large market affiliates to air only network commercials.

The format networks did not completely replace the tape-distributed syndication companies however. But the concept proved so successful that several other satellite-delivered specialized programming services were established. These new services focus primarily on music, sports, business, and financial news and include companies like Jones Satellite Network and the Business Radio Network.

To the average radio listener, there is little distinction between around-the-clock satellite programming and local radio programming. Proponents of satellite format networks say that there is no adverse or negative impact on local listening audiences.

During their first decade, however, satellite-delivered format stations in many markets registered lower total audience shares than stations with locally-produced programming. Since then, ratings have steadied and satellite formats have become a winning economic venture for many small markets that were financially strapped during the recession of the early 1990s.

The technology of satellite radio feeds also set the stage for new full service program entertainment networks, notably the RKO Radio Network in 1979 and the United Stations Radio Network in 1981. The principal founders and partners for United Stations included noted entertainer Dick Clark and former Mutual Broadcasting executive Nick Verbitsky. Both networks eventually offered multiple, demographically based program services.

Satellite technology created a large market for new program distributors by making it easier to establish *one-time* networks for live event broadcasting and ad hoc networks for short-form feature programming.

The ABC, CBS, and NBC radio networks readily embraced satellite technology for their existing traditional radio news networks while launching new, youth-oriented entertainment networks. They took advantage of a boom in discretionary income of 18- to 35-year-olds and a resurgence of the tight-sounding Top 40 AM radio format of the 1960s. That format, which became known as "contemporary hit radio" to a new generation of radio listeners in the early 1980s, became the rage of FM stations across the country. Traditional news networks, on the other hand, were attracting older listening audiences.

ABC pioneered the multiple (demographic) network concept in 1968, more than a decade before mass satellite distribution was introduced, when it formed four demographically targeted networks: ABC Information, ABC Entertainment, ABC Contemporary, and ABC-FM—the latter two capitalizing on the increasing number of FM contemporary music stations. The multiple network concept was implemented by then-ABC Radio Network president Walter Schwartz and ABC Radio Corporate Vice President Ralph Beaudin to address the growing variety of different radio formats at the time.

Among the youth-driven satellite networks launched by the major broadcast networks were The Source, created by NBC in 1979; RadioRadio, formed by CBS in 1982; and the Rock Radio Network, established by ABC that same year. All three networks distributed both special and regularly scheduled programs as well as news reports and so-called lifestyle or light news features. In addition, satellite technology allowed ABC and NBC to launch national two-way talk programming networks. (See Chapter 4.)

Existing radio program suppliers or syndicators also capitalized on the new satellite technology, which allowed them to more effectively compete with large networks in the distribution of specialized entertainment programming. Leading the pack of satellite program suppliers was Westwood One, a company that derived its name from its original base of operation in Westwood, California, a Los Angeles suburb. These networks and programming companies successfully brought many of the nation's more talented local radio personalities to the attention of radio listeners across the country.

> Satellites opened up opportunities for live broadcasts of programming for very reasonable costs. A company could become a radio network for an hour without having to put in permanent telephone lines to a permanent group of affiliates that was both very cumbersome and costly . . . . This changed the face of network radio.—Norm Pattiz, chairman of Westwood One[4]

As the number of new satellite networks and program suppliers multiplied, it became more difficult for radio networks to compete against each other and with other media for available advertising dollars. This was especially true for the large traditional or "line," as they are sometimes called, news networks. Network radio ownership began to contract.

Beginning in 1985, a wave of consolidations overtook the network radio industry when Westwood One, which made its mark in young-adult programming, purchased the Mutual Broadcasting System, a major news and sports network founded in 1934, from the Amway Corporation. Westwood One would later add the NBC Radio Network division, which included The Source, to its operations. With these acquisitions, Westwood One bridged the gap from pure program supplier to a full-service network radio company that now offered stations regularly scheduled news and feature services. Also in 1985, the United Stations Radio Network purchased the RKO Radio Network.

During a significant downturn in network radio's fortunes in the late 1980s, the Transtar Radio Network merged with the United Stations to form a new company called Unistar. At the same time, the ABC Radio Network, through a stock purchase, acquired control of the Satellite Music Network.

This consolidation mode continued into the early 1990s as Unistar, through a managing agreement, became part of Infinity Broadcasting—considered by many radio executives to be the most successful radio company to surface during the entrepreneurial eighties.[5] (See Chapters 3 and 5 for more on Infinity.)

Unistar was subsequently sold to Westwood One in 1994, making Westwood, along with ABC, the two largest network radio program producers in the United States. As part of a three-way deal, Infinity acquired an equity stake in Westwood One coupled with a management agreement. Unistar and Westwood were eventually combined under the Westwood umbrella with Infinity Broadcasting CEO Mel Karmazin assuming the same role for Westwood One.[6] (In 1994, Clark and Verbitsky again teamed up to form a new United Stations syndication company.)

Network consolidation and increasingly available satellite technology significantly expanded the number of network affiliates under single ownership by giving nearly all major networks multiple station affiliates in the same market. That included both primary or basic network affiliates and what are sometimes called "special program" affiliates—stations that only carry a network's select long-form music or talk programming.

Radio broadcasting lends itself more to multiple affiliations compared to television broadcasting because there are usually many more radio stations—with varying formats—serving each city. In some major cities, radio stations outnumber television stations (excluding cable services) three to one.

The solidification of the network radio marketplace, however, was not limited to the buying or merging of radio networks. As a competitive marketing tactic to attract more advertising dollars, major network companies consolidated their internal structure by further combining individual networks by demographics, a strategic sales effort to reach larger audience segments.

By the mid-1990s, the major network companies in radio were reduced to three general market organizations or forces: ABC, CBS, and Westwood One. Added to this mix was the merger of the Sheridan Broadcasting Network and National Black Network, both of which target the black radio market, into the American Urban Radio Network.

The acceptance of the new satellite format networks by cost-conscious radio station owners, especially following the elimination of public affairs programming requirements, also contributed to a downsizing around the country of local radio news staffs. In the FCC's early 1980s radio deregulation order, the commission said it was convinced that "significant amounts of non-entertainment of a variety of types will continue on radio." The FCC believed it no longer had to police this programming due to the growth of radio and other informational and entertainment services. The FCC further stated that the "only non-statutory programming obligation of a radio broadcaster should be to discuss issues of concern to the community of license."

Prior to deregulation, the FCC required that a certain percentage of a station's broadcast week be devoted to public affairs programming. Deregulation did away with those guidelines. Instead, stations now have to provide lists of local public interest needs in their public information file and document how their programming responds to those needs.

In the years following the FCC's initial deregulation order, several industrywide programming surveys showed that the amount of time allocated by stations to news was dropping down. Radio news executives at many music stations were at odds with management who cut news budgets as a cost-containment measure. According to a 1992 *Radio & Records* (a weekly trade magazine) survey of seven hundred radio stations, many music stations air shorter newscasts—two minutes or less—in fewer timeslots than in previous years.[7]

Music radio news executives were directly affected by growing information options for their listening audience including the increasing popularity of Cable News Network (CNN) and the development of news/information/talk programming on AM stations.

Local Versus National

Aside from news, there is a much larger programming issue looming in the background. While radio executives note that radio derives its strength from being a locally oriented medium—75 to 80 percent of a radio station's annual revenue comes from local advertising—local radio broadcasting also means being a distribution/syndication system for the nationally delivered satellite formats and programs. The main question posed is how much of radio programming will be nationally controlled via satellite in the twenty-first century? (See Figure 1.3.)

The satellite issue also invites the following question: From an economic standpoint, will national satellite programming permeate radio beyond the small- and medium-size markets?

There is also a new twist occurring in morning programming. Over the years, radio stations have continually registered the largest share of their overall listening audience between the hours of 6 A.M. and 10 A.M., Monday through Friday, when listeners actively listen to radio for local information. Known as morning "drive-time," the success or failure of the morning show can frequently set the tone and direction for a station's entire programming schedule—greatly impacting its advertising revenues. This is why some 24-hour satellite network affiliates prefer to keep their morning drive-time slot locally produced, thereby giving station management greater control over content.

The unyielding ratings successes of controversial morning talk show hosts Howard Stern and Don Imus prompted Infinity Broad-

Radio Programming Life Cycle

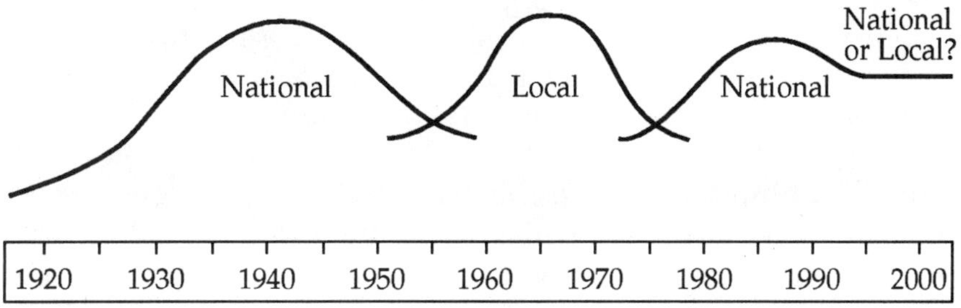

FIGURE 1.3

casting to nationally syndicate these local New York shows—first Stern, then Imus—by satellite to other cities. Both Stern and Imus on WXRK-AM New York and WFAN-AM New York (both owned and operated by Infinity Broadcasting), respectively, successfully compete with local radio talent in these markets.

Although their programs can have a New York slant, each with its own style, both Stern and Imus have transcended their local roots. Their at times topical, but always no-holds-barred approach to discussing issues has received national attention in major magazines and on television network news programs. In Imus's case, for example, he has amassed a following among noted political leaders and has interviewed many of them, including Bill Clinton, both as a candidate and as president.

While personality-based shows have been distributed by radio networks for some time, particularly during the 1980s, the move to distribute local shows designed for AM and FM morning drive-times on a national or regional basis is relatively new. What has developed is a series of daytime *SyndiNets*. (See Chapters 4 and 5 for more on talk programming.)

The trend in morning drive syndication lends credence to the notion that radio programming is slowly returning to the way it was in the thirties and forties—moving from locally to nationally based programming.

FM POPULARITY

For most FM radio stations, particularly those programming contemporary rock music, competitive morning ratings were not fully achieved until the 1980s.

During morning drive-time slots, the successful formula used by many FM stations involves a central personality or personalities—generally comedic in nature—surrounded by an ensemble cast composed of separate weather, traffic, and news announcers, all of whom interact during the broadcast. The mission of morning radio talent is to create an entertaining environment in which the audience experiences their "fun." This works for afternoon drive-time periods as well. Some AM stations also employ this ensemble approach in programming.

But to more fully understand FM radio's popularity among listeners, we need to backtrack for a moment. FM radio, which uses a wider frequency bandwidth than AM for transmitting signals, was developed by radio broadcasting pioneer Major Edwin H. Armstrong during the early 1930s. The wider bandwidth allows for a truer reproduction of musical sounds.[8] Since FM occupies a higher portion of the radio spectrum, there is less static and noise interference than at lower frequencies. AM signals, on the other hand, incur a tremendous amount of noise interference due to the propagation curves of AM waves.

It wasn't until the late 1960s that FM took on a personality of its own when stereo programming or audio transmitted and received over dual audio channels for greater fidelity (authorized by the FCC in 1961 to foster early FM growth) began attracting the youth-oriented contemporary music formats that dominated the AM band. At the same time, FM's stereo capability encouraged a new rock format that featured longer album selections that did not fit into AM's faster-paced, controlled pop music style. This new format was coined *progressive rock* by industry programmers and trade press. It later evolved into the present-day *album-oriented-rock* (AOR) and *adult rock* formats. FM stations also typically aired fewer hourly commercial spots than AM, which further induced AM radio listeners to turn their dials to the newer band.

On the receiving end, FM stereo radios became the passion of automotive manufacturers, who were then targeting young-adult "mobile" consumers who spent a great deal of their time listening to the new rock stations. As a result, AM stations lost a sizeable portion of audience to their FM counterparts as most contemporary music programming eventually left the AM band.

Beginning in the early 1970s and extending through the 1980s, the number of new FM stations signing on the air substantially outdistanced new AM stations. The AM radio band soon became associated with the demographically older "news and information" and "talk radio" formats. FM radio had arrived.

AM stations, however, still compose a large part of the affiliation base for traditional radio news networks, which provide AM stations with a full array of programming services, including veteran network news personalities Charles Osgood (CBS) and Paul Harvey (ABC). Additionally, during the 1980s CBS invested heavily in sports programming, while ABC, NBC, and Mutual launched national telephone-talk shows for their AM affiliates. (Chapter 4 for contains more on AM radio and its audience relationship to FM).

A CHANGING OF THE GUARD

Besides the enormous financial and technological changes impacting radio, several major acquisitions affected the broadcasting industry at large. With increasing deregulation, broadcasting was now a more desirable business, presenting financially viable opportunities for mergers and takeovers. A period of swift megamergers and takeover attempts was triggered in March 1985 with an announcement that shook the entire media world.

Capital Cities, a medium-sized, financially lean broadcasting concern principally involved in station operations, merged with (effectively acquiring) ABC, one of the United States' major communications conglomerates, for $3.5 billion. It was a good fit for both parties, especially considering the major market broadcast properties each brought to the table. The deal represented the largest media transaction of its kind to date. The American broadcasting structure that was in place for so many years was gone.[9]

In addition to being the first media deal of some consequence, the Capital Cities/ABC deal changed the ownership picture in radio by paving the way for government relaxation of broadcast cross-ownership rules—the so-called "one-to-a-market" rule. Adopted by the FCC in 1970, this rule prohibited new, common ownership of radio and television stations in the same market. (A commonly owned AM/FM station serving the same market is permitted by the FCC.)

Under this rule, an entity that owned a television property could not purchase a radio station (AM, FM, AM/FM, or as we will see later, AM/AM or FM/FM) in the same market or city where that station was licensed and vice versa. Additionally, existing radio-television combinations in the same market could not change ownership without selling either the radio or television property.

Capital Cities' purchase of ABC gave the newly configured media conglomerate radio and television combinations in several key markets: New York, Los Angeles, Chicago, and San Francisco. Some radio stations were spun off as a result of the merger. But in July 1989, the FCC granted Captial Cities/ABC a permanent waiver to its "one-to-a-market" rule, allowing the new company to maintain radio and television stations in those four markets, all of which were owned by ABC.[10]

ABC had been "grandfathered" under the 1970 FCC crossownership rule. That is, it had owned and could continue to operate its same market radio-television combinations when this rule was instituted. But as discussed above, the change in ownership made ABC's special status obsolete.

In effect, what the FCC did was adopt a waiver policy to its "one-to-a-market" rule for the entire broadcasting industry. The FCC said it would "look favorably" upon waiver requests in transactions involving television-radio combinations (TV-AM-FM) in the top 25 markets that have at least thirty separately-owned broadcast licensees or "voices." The four Capital Cities/ABC markets covered by the waiver were in the top 25 markets. (The FCC may soon ease the TV-radio crossownership rule further with a new television deregulation ruling.)

ABC was not the only media concern where major changes were underway. General Electric was readying its reentry into the multi-billion-dollar broadcasting ownership business. In December 1985, GE engineered the purchase of NBC's owner, RCA, for $6.3 billion.[11] The price substantially surpassed Capital Cities' acquisition of ABC the previous March.

General Electric was one of the three founding partners, with RCA and Westinghouse, of the NBC Radio Network in 1926, making it the world's oldest radio network. But RCA bought them out in the early 1930s and remained NBC's sole owner through 1985. General Electric, however, had been operating a separate radio station group until 1982.

Meanwhile, the inner sanctum of the broadcasting industry's most conservative network, CBS, was being disrupted by cable television mogul Ted Turner. During the spring of 1985, Turner attempted a hostile takeover (a change in the controlling interest of a company other than by acquisition) of CBS through a major stock offer valued at approximately $5 billion. The deal would have left Turner with a controlling interest in CBS.[12]

Turner was anxious to expand his television empire, which then included the Turner Broadcasting System, superstation WTBS-TV Atlanta, which is delivered via cable to various cities around the United States, and the Cable News Network (CNN). In order to be a leading broadcasting competitor, Turner needed access to more television households. CBS would give Turner that access.

To fend off Turner's takeover attempt and potential future takeovers, CBS embarked on a recapitalization program by buying back 20 percent of its stock. (Generally, recapitalization can take the form of refinancing, new equity infusion, or stock buybacks—the latter making corporate takeovers more difficult.) Then in October 1985, CBS invited Lawrence Tisch, then chairman and chief executive officer of Loew's Corporation, a New York City–based conglomerate that had substantial ownership in CBS, to join its board of directors. That ownership stake eventually went from 11.7 percent to 24.9 percent.

In September 1986, Tisch became acting chief executive officer for CBS and soon settled into the role of CBS president. CBS founder and past chairman, William Paley, became acting chairman. (Paley died in October 1990.) Under Tisch, CBS eventually sold two key assets—its record and publishing divisions—in order to build additional cash reserves.

This period of rapid-fire ownership and structural changes among the nation's top broadcasting companies was followed by several broad economic changes in the marketplace, including a marked decline in automotive and retail sales, two major sources of broadcast advertising revenues, coinciding with a reduction in consumer spending. These changes caused a *ripple effect* throughout the broadcasting industry. As advertising dollars became extremely tight, so did the media economy.

In addition, network television production and operating costs were soaring. The situation induced newly installed management at the large network companies to institute stringent fiscal restraints that gradually filtered down to their respective radio divisions. The changing economy also prompted the networks to explore potential new profit-making ventures.

Those factors were particularly evident at GE, which was fighting to keep NBC in its newly conquered first place ratings position. General Electric, deciding to focus its full attention on the television business, in 1986 opted to invest in cable and sell its radio division (networks and stations).

General Electric nearly struck a deal with Westinghouse Broadcasting in September 1986 to merge the NBC radio division with Westinghouse's radio division, creating a new, highly influential radio broadcasting company. Both GE and Westinghouse initially would have retained some of the new company's stock. But the deal fell through that December.[13] The following year General Electric sold the NBC radio station group to Indianapolis-based Emmis Broadcasting, one of the new entrepreneurial radio groups of the 1980s, and, as previously noted, its radio network division to Westwood One, which retained the NBC name.[14]

So began the rise of the nouveau radio station groups and networks. These new companies operated with less overhead compared to the major media conglomerates, a major financial attraction for investors. Through this structural change, radio broadcasting quickly became a more management-intensive business.

THE RISE OF THE MEGA-REPS AND MEGA-AGENCIES

A trend toward multiple national advertising representation of radio stations in the same market—that is, one company representing the interests of multiple stations in a given market—combined with a series of mergers and acquisitions among some leading advertising agencies, led to a major consolidation within radio's representation industry.

Basically, the role of a national representation firm is to secure national advertisers, which typically are major corporations and organizations represented by either advertising agencies or media buying services, for advertising on radio stations in cities around the country.

These accounts place national "spot" advertising on a select group of radio stations on a market-by-market basis. The stations can be grouped by ratings, format, and region of the country, as well as by demographics and listener characteristics. This differs from a conventional radio network buy whereby a purchase usually entails all, or most, station affiliates of the network.

Radio stations sign contractual agreements with representation companies to represent them to agencies and media buying services for advertising time sales.

In effect, the radio representation firm functions as a middleman in the national advertising selling and buying process between local radio stations and agencies or advertisers. In a marketing structure that has been in place since radio's early era, the rep company typically receives a 15 percent commission for every advertising dollar it places on a client station. The station receives the remaining 85 percent. (See Figure 1.4.)

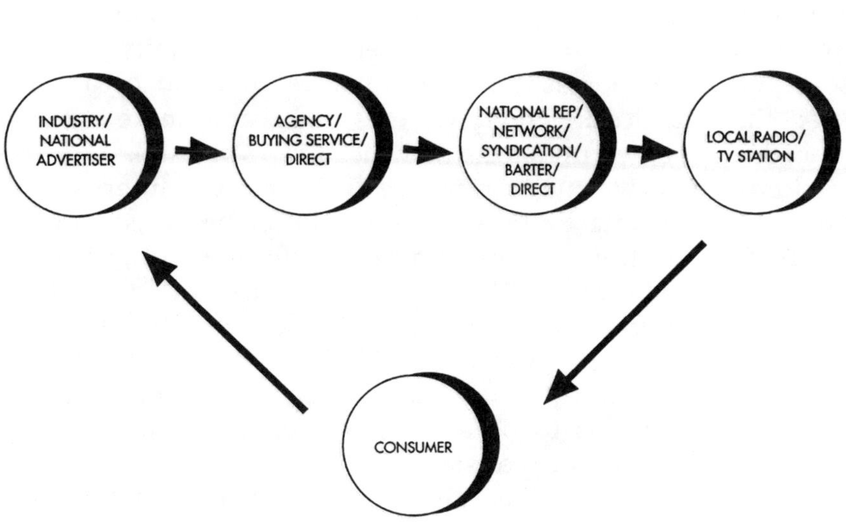

FIGURE 1.4. National market distribution model for radio

By their close work with the advertising community, the representation company's performance in obtaining national advertising dollars helps to attract top-rated stations to its client station roster. In some cases, national representation companies find themselves competing with national radio networks for a particular account or buy.

Before the 1980s, many independent national radio representation firms existed alongside major group-owned or "in-house" companies such as ABC, NBC, CBS, and Group W (Westinghouse Broadcasting) Radio. During the 1960s, for example, there were nearly forty individually or group-owned national radio representation firms. But that figure eventually dwindled to a handful.

Many group-owned representation companies required a large amount of operating cash while they consumed management time and resources. These rep entities soon became unprofitable subsidiaries and were eliminated. Their station representation lists went to the larger, independent radio rep companies. These same large rep companies either bought the remaining smaller independent firms or established new ones during radio's decade of mergers and acquisitions in the eighties.

What developed was the "mega-rep"—separately managed, individual representation firms that maintain autonomy but share common ownership. National representation firms under one corporate entity were now representing two or more stations in the same market to the same advertising community. Although some national representation companies in radio occasionally used this practice in previous years, it became more commonplace and accepted under the new mega-rep structure.

Radio has two major rep organizations: the Interep Radio Store and the Katz Radio Group, the latter being part of Katz Media Corporation, a radio and television marketing firm. The Interep Radio Store actually evolved from the McGavren Guild Radio firm, founded as the Daren F. McGavren Company in the early fifties. In 1981 McGavren Guild Radio established the Interep Radio Store (then called Interep) as the holding company for what was to be a series of rep company start-ups, acquisitions, and an expansion of services. With the formation of Hillier, Newmark, Wechsler & Howard, the Interep Radio Store concept began with an expansion of companies and services to position itself against an anticipated consolidation of buying power among most major U.S. advertising agencies. (Hillier, Newmark, Wechsler & Howard has since merged with another Interep company, Durpetti & Associates, to form D & R Radio.)

During the mid-1980s, Katz Radio, along with its parent (then known as Katz Communications) formed the Katz Radio Group and began buying national radio representation companies. Their biggest acquisition was Blair Radio (now called Banner Radio) in 1987. Blair Radio was one of the oldest radio representation companies in the business. The Katz Radio Group subsequently emerged as the radio industry's other mega-rep organization. Radio's national selling structure—traditional network and national spot business—had been consolidated by 1990.

Meanwhile, the merger frenzy infiltrated the advertising agency arena. The U.S. agency mergers were partially a response to the acquisition activity and multinational growth among corporations or agency clients, and partially a response to increased operating costs. As clients got bigger, so did the agencies. Corporate ownership for some major agencies or "shops," as they are called, had shifted from the United States to England.

The first consolidation move of major consequence during this time took place in 1986 when Saatchi & Saatchi P.L.C., a London-based advertising agency established in 1970 by two brothers, Charles and Maurice Saatchi, aggressively pursued and acquired, for approximately $450 million, Ted Bates Worldwide, a major New York–based agency. That price set a record for an agency acquisition.[15]

The deal catapulted the Saatchi brothers, who followed a growth-through-acquisition strategy to the top of the advertising world ranks. Saatchi & Saatchi quickly became a household name in the international media community. Ted Bates eventually merged with Backer & Spielvogel, another Saatchi & Saatchi U.S. acquisition, to form Backer, Spielvogel & Bates.

The catalyst for the Saatchi brothers' purchase of Ted Bates is believed to be the merger of three major U.S. agencies: BBDO International, Doyle Dane Bernbach, and Needham Harper Worldwide, under a single holding company called Omnicom.[16] (The latter two agencies have since merged.)

In 1987 the agency world witnessed the arrival of another British mega-organization. Martin Sorrell, the former financial director of Saatchi & Saatchi and owner of a two-year-old company that primarily made wire shopping baskets, Wire Products & Baskets (WPP Group P.L.C.), acquired J. Walter Thompson, one of the world's oldest agencies, for $565 million. Two years later, Sorrell's company bought Ogilvy & Mather for a substantial price of $865 million.[17]

With the addition of these two U.S. advertising agency powerhouses and several other media companies WPP already operated, Sorrell's relatively small British firm was now competing directly with Saatchi & Saatchi. The influence of foreign-controlled agencies upon the U.S. advertising world, however, had just begun. By 1990, Dentsu, a Japanese-owned agency based in Tokyo, had established itself as a top international marketing and advertising concern. Agencies in France also expanded their portfolios with ownership stakes in several U.S. advertising companies, including Foote, Cone & Belding (FCB Communications), and Wells, Rich, Greene.

Even though agencies under common ownership strove to operate independently of each other, rival advertisers soon found themselves conducting business within the same agency conglomerate, leading to discord among some clients. This eventually impacted agency fortunes as many major accounts switched shops.

By the beginning of the 1990s, advertising agencies had become the quintessential middle agent of the faltering media economies as they fought advertising budget cutbacks by their existing clients. The same tight operating economies hitting the broadcasting industry now strangled agency markets. Many large firms were besieged by debt. To rebuild their financial base, agencies had to lay off staff and sell corporate subsidiaries.

LEVELING THE FIELD

The concept of a holding company owning and operating separate advertising agencies is not new. Two decades before the mega-merger mania of the 1980s (1961 to be exact) the Interpublic Group of Companies was established by the McCann-Erickson agency so its management could operate more than one agency. In fact, the Interep Radio Store's multiple rep company concept is borrowed from Interpublic. The major agencies under the Interpublic umbrella include McCann-Erickson Worldwide and Lintas Worldwide.

The mega-rep structure in radio has put national representation companies on a level playing field with the large agency conglomerates. The mega-rep philosophy maintains that by using the combined resources of a large rep organization, more major national accounts will be acquired, causing additional advertising dollars to be spent on radio. Similarly, the merged network radio companies believe that their various consolidation moves have given them the added services required to deal with the larger agencies.

With these changing forces at work, the radio industry still was able to support one group-owned broadcast representation company: CBS Radio Representative, among the oldest of the in-house radio representation companies. Primarily, it represents all of the CBS-owned and -operated radio stations to advertisers.

There are also several smaller, privately owned national and regional representation companies that essentially represent smaller market radio stations. But by the 1990s, the Interep Radio Store and the Katz Radio Group divided, nearly equally, about 90 percent of the total national spot advertising on radio.

A NEW OLD INDUSTRY

Today, the United States' oldest electronic communications medium is a new and different industry with new faces, new attitudes, and new challenges. At a time when listeners and advertisers alike can choose among an increasing number of media alternatives, many entrepreneurial radio group operators have a renewed determination to position radio in the communications forefront.

In essence, radio's relationship with society has come full circle since its beginnings. Radio started out being the primary national news, music, and variety outlet of the twenties, thirties, and forties. With television's arrival, radio evolved into a local medium, serving as the main source for rock 'n roll music and the resulting lifestyle and cultural changes of the fifties and sixties. As it became a more specialized medium in the seventies, radio countered somewhat the rise of network television's popularity. And in the eighties and nineties, the medium experienced total rebirth as a powerful information and entertainment force in the "high-tech" age.

SUMMARY

The emergence of this new radio industry was primarily fueled by the following factors:

- *Federal deregulation of the radio broadcasting industry* by the FCC beginning in 1981 as it pertained to rules and policies governing programming and ownership
- *The FCC's elimination of its three-year anti-trafficking rule,* which prohibited station owners from selling their properties within a three-year period. The rule was originally adopted by the FCC in 1962 to foster continuous local community broadcast service. Elimination of the rule allowed a radio station's license to become more marketable. In the mid-1980s, there was also a *repeal of the FCC's national ownership cap on radio stations from seven AM and seven FM stations to twelve on each band—fourteen on each band for minority-owned stations.* But it was the abolition of the three-year rule by the FCC that paved the way for an influx of financial (capital and equity) funds from lending institutions for investing in commercial radio.
- *The introduction of satellite technology for national program distribution* through the launch of satellites by various telecommunications companies, including Western Union and RCA (now General Electric), in the late 1970s and early 1980s. Satellite transmissions allow simultaneous distribution of nationally produced programs from a central source anywhere in the country. Previously, radio stations received programming signals from a network over land (telephone) lines—one program at a time—or through the mail, as was the case for specials or weekly series produced on either records or tapes—a distribution form that still exists. Satellite technology permitted radio

stations to more easily choose what they preferred to broadcast live than in past decades. It also provided stations with better network sound quality, cost savings over local production, and easier access to multiple network affiliations.

- *The growth of FM radio* during the 1970s and 1980s, which can be characterized as FM's coming of age. The number of new commercial FM stations far exceeded that of AM. This initial growth rate proved that FM radio as an entertainment medium was in demand by the public, leading to increased advertising revenues.
- *The balance of radio buying and selling power.* The concentration of buying power through a large number of agency mergers and acquisitions was offset by the consolidation and eventual formation of major network radio forces: ABC, CBS, Unistar/ Westwood One, the American Urban Radio Network; and two major national representation forces: the Interep Radio Store and the Katz Radio Group, giving radio the necessary resources and clout to be effectively marketed on a national basis.

NOTES

1. *Broadcasting: The First 50 Years,* Broadcasting Publications, Inc., 1982, p. 246; *The Broadcast Century,* (Focal Press, 1992), 233–34.
2. "Court Upholds FCC's Striking of Three-Year Rule," *Broadcasting Magazine,* Sept. 3, 1990, pp. 50+.
3. Mark Reynolds Dodd, *Satellites & Radio Broadcasting: A Historical Review and Market Developments,* Marcia L. DeSonne, ed., National Association of Broadcasters, p. 3.
4. Based on a conversation with Norm Pattiz, chairman of the board and former president of Westwood One, Culver City, Calif.
5. "New Network Giant: Infinity to Run Westwood and Unistar," *Broadcasting Magazine,* Oct. 18, 1993, pp. 40–41; "Westwood One Restructures Into Two Divisions," *Radio & Records,* April 22, 1994, pp. 1+.
6. Ibid.
7. "R & R Survey Probes News on Music Radio in the '90's," *Radio & Records,* Nov. 27, 1993, pp. 1+.
8. George H. Douglas, *The Early Days of Radio Broadcasting,* (Jefferson City, NC: McFarland and Company, 1987).
9. "CapCities + ABC: A Bold $3.5 Billion Media Marriage Electrifies Industry and Nation," *Broadcasting Magazine,* March 25, 1985, pp. 31–36.

10. "FCC Grants CapCities/ABC Permanent Waiver," *Radio & Records*, July 7, 1989, pp 1+; "ABC Can Keep Radio Stations," *Broadcasting Magazine*, July 10, 1989, pp. 29–30. NOTE: Where there were multiple, same-market Capital Cities and ABC-AM and -FM properties, several of these stations were sold off to meet the then FCC duopoly requirement of owning one AM and one FM station in a given market.
11. "RCA + GE: Marriage in Takeover Heaven"; "NBC Greets GE with Open Arms"; "RCA & NBC: The Way They Were"; "Merger Reaction: Hopes for An Unchanged Course," *Broadcasting Magazine*, Dec. 16, 1985, pp. 43–48.
12. "Ted Turner's Bold Bid for CBS"; "What the Rest of the World Thinks of Turner/CBS"; "Ted Turner: Maverick and Media Missionary"; "CBS Begins to Fight Back," *Broadcasting Magazine*, April 22, 1985, pp. 30–35.
13. "CBS's $1 Billion Counterattack," *Broadcasting Magazine*, July 8, 1985, pp. 31–32; "Turner Blasts CBS Management;" "Turner in Trouble?" *Broadcasting Magazine*, July 15, 1985, pp. 30–32; "FCC's Double Whammy to Turner's CBS Hopes," *Broadcasting Magazine*, Aug. 5, 1985, p. 30; "CBS Takes Out an Insurance Policy," *Broadcasting Magazine*, Oct. 21, 1985, p. 36.
14. "Group W, NBC Radio Marriage Is Off," *Broadcasting Magazine*, Dec. 16, 1986, p. 54.
15. "Bates Oks Saatchi Talks," *Advertising Age*, May 5, 1986, pp. 1+; "S&S $450 Million Buys Bates," *Advertising Age*, May 12, 1986, pp. 1+.
16. "Biggest Global Agency: DDB–Needham–BBDO," *Advertising Age*, April 28, 1986, pp. 1+.
17. "JWT Group Sold," *Advertising Age*, June 29, 1987, pp. 1, 62; "Ogilvy, WPP Talking Deal," *Advertising Age*, May 15, 1989, pp. 1+; "Ogilvy Under WPP Wing," *Advertising Age*, May 27, 1989, pp. 1+.

2

Radio's Fiscal Dynamics

Before reading about the structural changes that have altered the radio broadcasting landscape, it is important to understand the fiscal characteristics of commercial radio.

THE INCOME AND THE RATINGS

A radio station's income is derived primarily from the sale of station airtime to local, national, and regional advertisers, usually through advertising agencies and national representation firms. (As noted previously, local advertisers account for a large portion of a station's advertising revenue.) Additionally, stations receive paid compensation from major radio networks for the clearance of their commercial spots. (Income also includes political advertising campaigns, trade-outs, and barter arrangements with advertisers.)

The amount a radio station charges for its advertising time depends upon its ratings or audience size, which can vary with the type of programming, time of day, and the audience's demographics.

Arbitron is the principal ratings and research company for local radio audience measurement, supplying audience data for five dayparts covering a 24-hour period. They have been measuring radio listening since 1964.

Arbitron uses a personal, seven-day diary methodology to record radio listening habits in over 260 U.S. cities. In many of these markets, audience listening data is offered quarterly by season—winter, spring, summer, and fall—and on a continuous or monthly (two-month rolling average) basis. Some smaller markets have only one or two yearly survey periods.

Arbitron presents its data as Average Quarter Hour (AQH) persons and cume (cumulative) persons. Listening estimates are provided for a variety of demographic age groups, starting with age 12. The AQH identifies the average number of persons estimated to have listened to a station for a minimum of five minutes during any quarter hour in a specific period.

AQH has become the basic buying and selling unit for determining a station's rating (percent of the population being measured), and share (percent of total radio listening by the population being measured). Cume reflects the number of different or unduplicated people listening during a specific period.

The AQH estimate helps determine both the size of audience and the cost of a spot schedule rotating within a time period or daypart.[1]

Most major agencies and buying services use Arbitron as the primary radio research service for planning a radio buy.

Arbitron also markets qualitative product purchase and lifestyle information. During the 1980s, lifestyle audience information for local radio stations emerged as a major image positioning and sales tool to differentiate stations from competitors and other forms of advertiser-supported media. Many research companies now compile general qualitative, product usage, and customized audience perception data for radio stations.

In 1992, Strategic Radio Research, a Chicago-based media research company, launched a customized local market radio ratings service called AccuRatings as an alternative to Arbitron. AccuRatings uses telephone interviews to obtain both qualitative and quantitative information.

The major difference between the two services is that AccuRatings computes daypart listening by a basic "station partisan" or preference share (most-listened-to station) gathered from its telephone survey, instead of the AQH breakouts used by Arbitron.[2]

Both ratings services present station listening levels for the entire day and by various time periods. As previously mentioned, radio stations typically attract the highest number of listeners during morning drive-time (6:00 A.M. to 10:00 A.M.), followed by late afternoon drive-time (3:00 P.M. to 7:00 P.M.), and midday (10:00 A.M. to 3:00 P.M.). Advertisers can buy time during specific dayparts or a combination of dayparts, known as run-of-schedule (ROS).

Audience measurements are also conducted to determine the number of radio listeners outside the home—in cars or elsewhere. According to Statistical Research, Inc., a Westfield, New Jersey–based media research firm that tracks radio audience and network radio listenership through its semiannual RADAR audience survey reports, there is more out-of-home than in-home listening occurring during the mobile nineties. (See Figure 2.1.)

Since the U.S. government's initial deregulation of radio in the early 1980s, music stations have maintained an average hourly com-

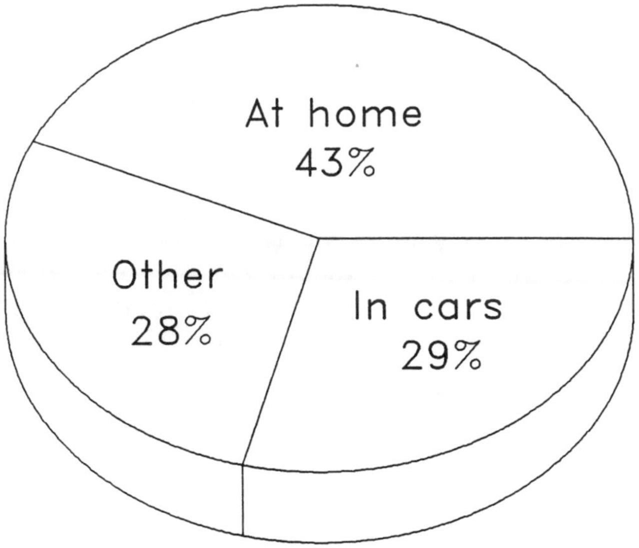

FIGURE 2.1. Share of radio audience age 12 and older by location (Monday– Sunday, 24 hours). "Other" includes those listeners who are shoping or at work. RADAR® fall 1993. Copyright© Statistical Research, Inc.

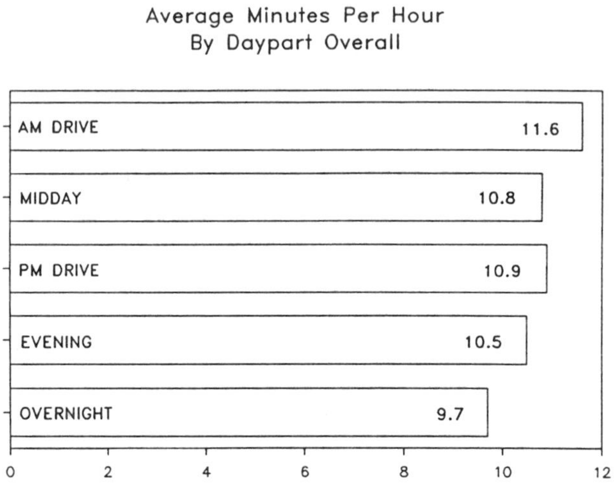

Market Size		
1–50	50–100	100+
11.1	11.6	12.1
10.1	10.9	11.5
10.1	11.1	11.5
9.9	10.8	10.8
9.6	9.7	9.8

FIGURE 2.2. Commercial loads. Source: *Radio & Records* Sales/Management Survey, 1991.

mercial load of 11 minutes. For all-news formats, the average can be slightly higher. This average hourly commercial "load," as it is called, is a positive marketing tool in that radio typically airs fewer commercials when compared to television and cable, meaning less commercial clutter for advertising messages. (See Figure 2.2.)

Income can also be derived from studio rentals and, as will be addressed later, subcarrier leasing and local programming and marketing agreements.

THE POWER RATIO

Radio stations try to achieve a high "power" or conversion ratio, which shows how well audience share is converted into market revenue share. In other words, the power ratio is used to evaluate a station's advertising income by examining its relationship to available radio market revenue. Basically, it represents a station's ability to convert its age 12 and over Arbitron audience share into comparable revenue shares.

Historically, news/talk, full-service MOR, and adult contemporary stations have shown the greatest ability to convert share into revenue, believed to be due to loyal audiences with higher than average disposable incomes. These rankings vary, however, according to individual markets.

Two firms that specialize in calculating various aspects of financial radio data, Duncan's American Radio, an Indianapolis-based radio research firm, and Miller, Kaplan, Arase & Co., a Los Angeles–based certified public accounting firm, both track power ratios by radio station and format. Each employs a different methodology, however. Generally, to determine power ratio a station must divide its audience share into its market revenue share.

If the ratio of converting audience share to revenue share (used to measure the effectiveness of a station's sales efforts and for showing potential market share revenues) is greater than one to one, the station is succeeding in receiving a larger piece of the radio revenue pie relative to its 12-and-older ratings. (See Chapter 6 for information on inventory management control.)

THE EXPENSES

To properly operate a typical radio station or group of stations, the owner(s) must set an operating budget, including salaries, in six principal areas: administration, advertising and promotion, engineering, programming and production, news, and sales.

An early 1990s survey of radio station finances jointly conducted by the National Association of Broadcasters (NAB) and the Broadcast-Cable Financial Management Association (BCFM), two industry trade groups, shows that on average, radio stations allocate the largest portion of their operating budget to administration (42.1%), followed by programming and production (21.1%), sales (19.7%), advertising and promotion (8.6%), news (4.3%), and engineering (4.1%) (See Figure 2.3.)

Budget allocations will vary depending on the type of radio station (AM, AM-FM, or FM), number of subscriptions to nation-

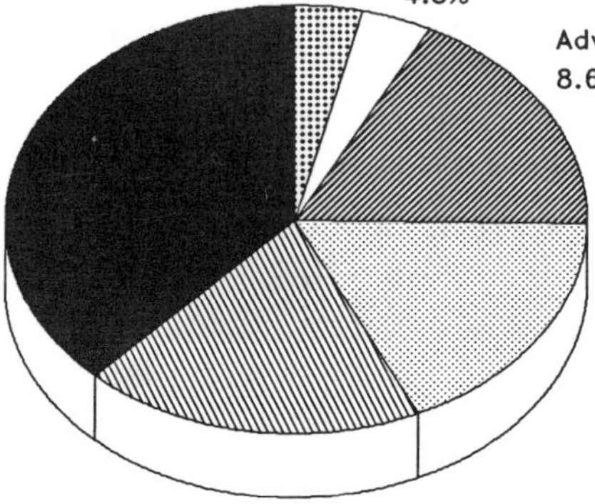

FIGURE 2.3. The national average breakdown of major operating expenses for a radio station. Source: *NAB/BCFM Radio Financial Report*, 1992.

ally delivered programming services, market size, and the overall competitiveness of the local media. Other factors that should be considered in determining budgets are the level of the station(s) income, the amount of expenses shared with other commonly owned stations, and current market conditions.

Additional expenses include music licensing fees—paid annually to music licensing organizations (passed on to artists, composers, and music publishers for the right to play their songs and theme music), ratings and research subscriber fees, consulting fees, and debt service payments to station investors/lenders.

AVERAGE SALARIES

As with most businesses, a large portion of a radio station's expenses goes directly to salaries. In an early 1990s salary and benefits survey of U.S. radio stations, the National Association of Broadcasters offers a general barometer of average base salaries for different staff positions at a typical local station. (See Figure 2.4.)

Annual salary figures for specific positions vary greatly, however, depending on station revenues, market demographics, and

ANNUAL AVERAGE BASE SALARIES FOR RADIO STATION PERSONNEL(IN DESCENDING ORDER)

Management, Non-Sales	*Sales Executives*	*Station Talent*
General Managers $56,458	National Sales Manager $70,983	Personalities/DJs $22,471
Program Directors $32,953	General Sales Manager $53,983	Sports Reporters $24,729
Operations Manager $30,275	Local Sales Manager $52,617	News Reporters $19,449
Chief Engineers $28,748	Account Executives $32,184	
Promotion Director $26,855		
Business Manager $25,632		
News Director $22,352		
Sports Directors $20,889		

Source: *NAB Radio Employee Compensation and Fringe Benefits Report*, 1992.

FIGURE 2.4. Annual average base salaries for radio

common multiple ownership in the same area. And if there is a high listening level for morning radio shows, salaries for local morning personalities will be considerably higher than those of the rest of a station's on-air talent.

The chart shows major radio station positions, including support staff. For sales personnel, the salary figures represent their total compensation, which can include any combination in salary, commission, and bonus. Important to note is that base salaries vary significantly between large markets like New York, Chicago, and Los Angeles and the nation's smaller cities. For instance, in a market where the population is over 2.5 million, the average salary for a program director is $75,263. However, in a market where the population falls between 500,000 and 1 million, the program director's average base salary drops to $38,363. And when the market population is between 50,000 and 100,000, the average salary drops to $20,254. Salaries also vary between AM and FM stations, with employees of the latter usually commanding higher compensation.

For non-sales station management in all size and type station markets, the average base salaries by job classification were (in descending order): general managers—$56,458; program directors—$32,953; operations managers—$30,275; chief engineers $28,748; promotion directors—$26,855; business managers— $25,632; news directors—$22,352; and sports directors—$20,889.

Looking at station talent, salaries for full-time on-air personalities averaged $22,471, while sports reporters and news reporters earned $24,729 and $19,449, respectively.

And on the all-important sales side, the NAB survey showed the following compensation averages by job classification: national sales manager—$70,394; general sales managers—$53,983; local sales managers—$52,617; and account executives—$32,184.

These general salary bases are generally lower than the corresponding positions in television. The average base salary for a U.S. television station general manager, for example, is $119,204, while the average salary for a station account executive is $48,671.[3] However, radio salaries continue to rise steadily.

When reviewing sales accountability for radio executives in the 1990s, it has become imperative for station owners and managers to know which activities lead directly to profits. According to an analysis prepared by Henry Lawson, chairman of the executive council for the Interep Radio Store, sales time for large market stations was more oriented toward agencies, while for smaller market stations sales time was more directed at client or advertiser decision makers. Station management must "determine the expandability of these revenue sources and allocate . . . resources accordingly," Lawson's report concluded.[4]

PASSING THE PROFIT THRESHOLD

With radio's performance primarily tied to seasonal ratings periods, which can fluctuate from quarter to quarter and sometimes month to month, a station's income statement may show marked revenue swings.

Since advertising rates closely follow ratings, it can be difficult for some stations to maintain a consistent commercial pricing structure. A basic pricing strategy based upon advertiser demand, however, is essential to maintain a strong earnings potential.

The major threat to a radio station's profitability is high fixed costs concurrent with declining advertising revenues. Since radio broadcasting is a seasonal business, stations should break through their profit threshold or that percentage of the station's

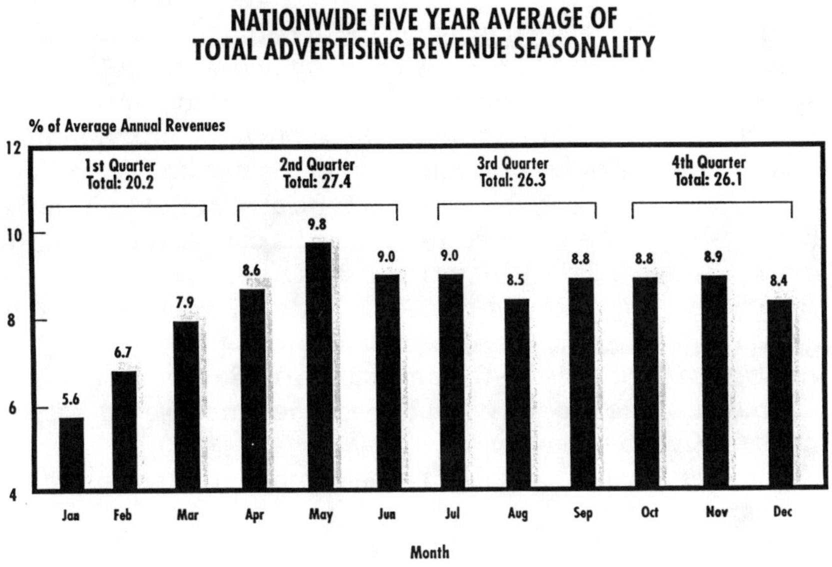

FIGURE 2.5. Nationwide five-year average of total advertising revenue seasonality. (Based on a five-year average from 1987 to 1992.) Source: Miller, Kaplan, Arase & Co., Certified Public Accountants.

projected annual revenues that will maintain a positive cash flow early in the year.

This fiscal approach is especially important to counter the traditionally low revenue-producing first quarter for radio. For a perspective on the seasonality aspect to the radio business, see Figure 2.5 for Miller, Kaplan, & Arase's five-year average monthly study of radio revenues.

DETERMINING VALUE

The principal criterion for determining a radio station's resale value is its *broadcast cash flow* (both realized and current)—operating income after taxes, plus noncash charges such as depreciation.

Other factors or criteria that apply to appraising the value of radio properties include

- *Market revenue growth expectation* (both national and local)
- *Revenue projections of property driven by market competitiveness and station position*—whether a proposed station acquisition is a mature or start-up property and how its programming relates to other radio programming in the market. Projections should also consider the station's typical share of total radio listeners (age 12 and older) as well as its core audience in the market and its advertising pricing structure.
- *Anticipated operating margins*—the ratio of pretax profit to gross revenue
- *Anticipated "exit value"*—the projected cash flow value a potential buyer of the property would base the purchase price upon in about five to seven years.
- *Cost of debt/capital*—calculating the combined interest rates of the various lenders.[5]

A THREE-TIERED FINANCING APPROACH TO RADIO OWNERSHIP

The potential radio station owner needs a sophisticated strategy in place before seeking financial assistance. In radio's ideal lending model, there are three different levels (or layers) of financing for acquiring existing radio broadcast properties.

The first level of financing involves the *senior lenders* or major banks. They use broadcast cash flow as the measurement for determining loans along with what they calculate as the "real value" of the property in the worst case (revenue and operating income) scenario. Basically, lenders assess the station's ability to generate cash that could then be available for debt service.

Senior lenders charge anywhere from the prime interest rate to prime plus three points. The second level is called *mezzanine* or *subordinated* lending where money is put up by institutional investors such as insurance companies and venture capitalists. These investments are expected to furnish returns beyond the prevailing prime interest rate.

Mezzanine lenders may hold the right to convert some portion of their investment into company stock or a significant ownership percentage of the business.

The third level of financing radio properties lies with the *equity* itself, also known as the "risk capital"—money that the buyer and/or the buyer's business partners place in the deal. For a radio group, equity can be derived from tapping the public market through an initial public offering (IPO). In other situations, equity can take the form of "venture capital" investor funding. This kind of financing may yield an even higher rate of return than the previous two tiers depending on the amount of risk assumed by the investors. The difference here is that typically there are no monthly debt service payments on venture capital.

Equity placement took on a greater financing role during the early 1990s due to a shortfall in major financing capital.

Foreign ownership is limited to only 25 percent of the capital stock of either a radio or television broadcasting property. This ownership cap was established by the U.S. government as a means to prohibit potential foreign influence over U.S. media.

TAX CERTIFICATES AND MINORITY OWNERSHIP

To foster minority ownership in the radio, television, and cable industries, the FCC allows tax certificates to be issued to those who sell to a minority-owned firm as a way to facilitate the capitalization of broadcasting and cable acquisitions by those groups. The FCC defines a "minority-owned" firm as one in which African Americans, Asian Americans, Hispanic Americans, and Native Americans own over 50 percent of a corporation's voting shares or 20 percent of the total equity in a limited partnership.

(Shareholders in a minority-owned broadcast or cable entity are also eligible for a tax certificate upon the sale of shares, provided their interest was purchased to assist in the acquisition's financing.)

Tax certificates allow both sellers and investors to defer the payment of capital gains tax on a broadcast or cable sale. One way to do this is to reinvest the sale's proceeds in a "qualified" replacement or similar property approved by the Internal Revenue Service. Generally, this means another media-related company.

Minority buyers can use this tax certificate policy, adopted by the FCC in 1978, to attract equity investors and negotiate for a reduction in the purchase price of the facility due to the tax savings accrued by the seller.[6]

An early 1990s statistical analysis of minority-owned commercial broadcast stations, conducted by the Minority Telecommunications Development Program of the National Telecommunications and Information Administration (NTIA), showed that of nearly ten thousand commercial AM and FM stations licensed in the United States, 2.8 percent were minority-owned. As part of a major 1992 radio ownership ruling, the FCC has proposed an "incubator" program for group owners who want to exceed the national station ownership limit by helping minorities and small businesses to become radio station owners. (See Chapter 5.)

A principal financial resource for minority buyers is the Washington, D.C.–based Broadcast Capital Fund, also known as BROADCAP. It is a private, nonprofit venture capital company funded by voluntary contributions from the major networks and other members of the broadcast industry. BROADCAP, which was formed in 1978, makes investments in both uncontested construction permits and existing broadcast properties.[7]

NOTES

1. *A Guide to Understanding and Using the Radio Audience Estimates,* the Arbitron Company, 1992, p. 1.
2. "AccuRatings: Challenging Arbitron and the AQH," *Radio & Records,* Oct. 30, 1992, pp. 34–35.
3. *Television Employee Compensation & Fringe Benefits Report,* National Association of Broadcasters, 1993.
4. "Determining the Price of Doing Business," *Radio & Records,* Dec. 31, 1993, p. 14.
5. Some of the information here and in Chapter 3 appears in a financial white paper entitled "Radio as a Long-Term Investment," prepared by the author and published by the Interep Radio Store in 1991.

6. *Minority Tax Certificate Sourcebook*, Minority Enterprise Program of the Federal Communications Commission, 1991–92.
7. Based on an interview with John Oxendine, president and chief executive officer of BROADCAP, Washington, D.C.

3

The Financial Boom and Radio's Emerging Entrepreneurs

> Entrepreneurs see change as the norm and as healthy. Usually, they do not bring about the change themselves. But—and this defines entrepreneurs and entrepreneurship—the entrepreneur always searches for change, responds to it, and exploits it as an opportunity.—Peter Drucker[1]

NOUVEAU RADIO

Our examination of America's entrepreneurial radio industry begins with a basic premise: Radio broadcasting is a product-driven business.

The backbone of the radio industry is the local radio station, which is in the business of selling its programming product to advertisers. The programming forms the basis of a station's image, causing listeners to either tune-in or tune-out. U.S. radio stations operate under a government license, renewable every seven years, to service their community with programming over a designated frequency—programming that is chosen by station management.

Because radio broadcasting entered a higher financial playing field during the 1980s with escalating sale prices, a results-driven economy was created. At times, this bottom-line mentality took precedence over programming concerns. Higher prices for station properties led to higher rate cards for advertisers and more revenue for the industry.

One yardstick for measuring the economic performance of the radio business is tracking annual sales. Although the radio industry garners only about 7 percent of all advertising dollars, overall radio broadcast revenues (network, national spot, and local radio sales) have increased annually in proportion to other media advertising expenditures.

From 1982, the beginning of radio deregulation, to 1990 the radio industry had a compound annual growth of 8.8 percent, climbing from $4.5 billion to nearly $8.9 billion in annual revenues. The increase demonstrates that radio was used more extensively by the advertising community. (See Figure 3.1.)

Radio industry advertising was also growing faster than inflation, as measured by the Consumer Price Index (CPI). From 1985 to 1990, for example, radio revenues had a composite growth rate of 34.5 percent, compared to the CPI's growth rate of 21.5 percent.

This steady growth, particularly evident in the mid-1980s, not only attracted major banks to radio, but also various financial institutional investors. What they saw was the potential for a high return on investment at fairly fixed costs. In short, these investors and lenders saw a business with the ability to maintain positive cash flow margins.

This is especially true for multiple station operations:

> If you own three stations and one of them comes under competitive attack and the other two stations are doing fine, the two will provide you with the cash flow necessary to meet all your (financial) objectives and shore up the third one. And over time, you can make adjustments.—Frank Osborn, a former RCA/NBC Radio executive who in 1985 formed Osborn Communications to acquire radio stations.[2]

The litany of major investors and lenders that had a part in the eighties radio broadcast acquisition frenzy included Morgan Stanley & Co., the New York–based stock brokerage firm that primarily arranges financial investments; CIGNA, the Hartford,

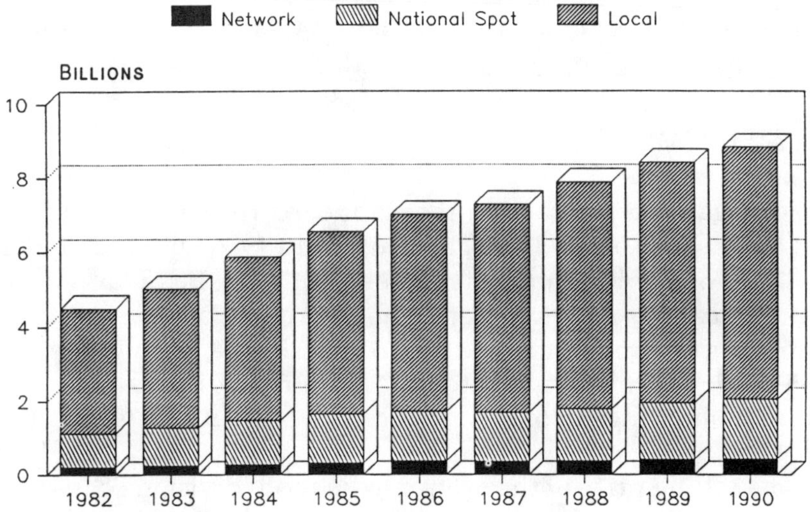

YEAR	NETWORK	NATIONAL SPOT	LOCAL	TOTAL
1982	$217.5	$909.4	$3365.0	$4491.9
1983	253.5	1022.8	3739.0	5015.3
1984	288.0	1184.4	4412.0	5884.4
1985	328.7	1319.4	4915.0	6563.1
1986	380.0	1332.6	5313.1	7025.7
1987	371.0	1315.0	5605.0	7292.0
1988	382.0	1402.0	6109.0	7893.0
1989	427.0	1530.0	6463.0	8420.0
1990	433.0	1626.0	6780.0	8839.0

FIGURE 3.1. Radio advertising expenditures in millions. Source: Radio Advertising Bureau.

Connecticut-based insurance company, which combined forces with Morgan Stanley to form the Morgan Stanley and CIGNA Leverage Capital Fund; and the equity investment house of TA Associates and the Bank of New England, both based in Boston.

Another attractive aspect of the radio broadcasting business to financiers is its defined marketplace, especially for most larger U.S. cities. This means that there are a relatively fixed number of radio stations in these markets, lessening the chances of an upstart company launching a new station. Significant value is placed on local radio stations that could reach specific audiences and, through specialized formats, could then become a radio programming franchise.

THE METROMEDIA DEAL

If there was one transaction that launched the soaring radio station trading era of the 1980s, it occurred in 1986 with the management buyout of the radio division of Metromedia, a large media communications conglomerate. The Metromedia radio group was sold by its chairman, media tycoon John Kluge, for a then record $285 million. Kluge wanted to divest Metromedia of all television and radio holdings to concentrate on the cellular communications industry.[3]

The selling price of the Metromedia radio division eclipsed the previous radio station group sale record by $178 million. In 1985, CBS purchased five radio stations from Taft Broadcasting (now known as Great American Broadcasting) for $107 million. In the Metromedia deal, Kluge sold nine of the company's ten radio station properties (six, including WNEW-AM/FM New York, were in the top 10 markets) along with the Texas State Networks, a regional radio network operation, to an investor group led by then Metromedia Radio president Carl Brazell; Morgan Stanley & Co.; and other members of Metromedia management. The sale resulted in the formation of a new radio company called Metropolitan Broadcasting.[4]

The amount of money involved in the Metromedia sale made both financial lenders and radio broadcasters stand up and take notice. The radio industry was inducted into a new level of business economics as station values soared.

ON THE BANDWAGON: OTHER SIGNIFICANT TRANSACTIONS

Other significant management buyouts in radio broadcasting occurred in 1986. There was the sale of Katz Broadcasting, the radio station unit of then Katz Communications, to a management group led by Dick Ferguson, then president of Katz Broadcasting. The new radio company became known as NewCity Communications. The radio broadcast division of Josephson International was purchased by its president Ed Christian and renamed Saga Communications. And Harte-Hanks Radio Group was bought by its president Gary Edens and renamed Edens Broadcasting.

This explosive amount of station trading was fueled further by the U.S. government's announced change in the tax code. Nineteen eighty-six was the last year that a broadcast property sale would incur only a 20 percent capital gains tax.

Additionally, after January 1, 1987, the government eliminated a tax rule, known as the general utilities rule, that protected management buyout transactions from double taxation, particularly those that involved distribution of funds to shareholders.

Because of these changes, some of these deals, depending on the organizational structure of the companies involved, would be subject to a new aggregate tax rate of almost 50 percent. The tax law amendment, however, did not hinder record station sale prices for long. Buoyed by the availability of easy financing, repeated station trading continued unhindered.

When it came to high-finance deal-making in radio in the eighties, Robert F. X. Sillerman, redefined the dimensions of radio station investing.

Sillerman first entered radio ownership in 1978 when he established a medium market broadcasting (radio and television) group with New York radio personality Bruce "Cousin Brucie" Morrow. That company dissolved in early 1985. In September 1985, Sillerman, along with investors William Magee and Howard Tytel, formed Sillerman-Magee Communications Management, a New York-based investment banking firm specializing in communications.[5] That November, Sillerman began to mold an aggressive image in radio deal-making when he acquired four radio stations from Doubleday (which was dismantling its radio broadcasting division) and one station from Metromedia.

Shortly after that Sillerman cofounded a new radio station group with Carl Hirsch, the former president of the Malrite Communications radio division (a Cleveland-based broadcasting concern that has since merged with Disney's Shamrock Broadcasting), who had just acquired KJOI-FM Los Angeles—one of that city's top-rated radio stations. These six stations formed the nucleus of a new company called Legacy Broadcasting. Sillerman and Hirsch were cochairmen, with Hirsch handling the day-to-day operations of the company.[6]

Sillerman, however, did not limit his radio portfolio to Legacy. In March 1988, during the National Association of Broadcasters (NAB) annual convention in Las Vegas, he unveiled plans for another record radio station deal involving the old Metromedia radio division: the acquisition of Metropolitan Broadcasting for $300 million. By this time Metropolitan had already sold two of the nine stations (WASH-FM Washington, D.C., and WIP-AM Philadelphia) from the original Metromedia purchase to trim its debt service.[7]

Sillerman-Magee Communications Management purchased Metropolitan, with Legacy Broadcasting retaining a near 80 percent interest in the group. With the addition of Metropolitan, Sillerman had assumed more than $200 million in public and private debt. To alleviate some of these liabilities, four stations from two radio groups—Legacy and Metropolitan—along with Metropolitan's Texas State Networks, were sold to former Metropolitan chief Carl Brazell for $145 million. This transaction formed the basis for a new radio company called Command Communications, headed by Brazell. The financial structure of Command Communications was managed by Sillerman-Magee.[8]

The Metropolitan and Legacy stations were eventually acquired by Westinghouse Broadcasting (Group W Radio) in 1989, making Westinghouse one of the largest non-network–owned radio groups in the United States.[9] The Westinghouse deal included FM radio stations in New York, Philadelphia, and Los Angeles, long-time AM radio markets for Group W. Group W Radio now had a potent audience market for advertisers with AM/FM properties in those major cities.

The move of Metropolitan and Legacy into Group W Radio, however, was not the end of Sillerman's radio holdings. Sillerman maintained radio station investments and formed a new group, SFX Broadcasting. Meanwhile, Legacy's Hirsch launched a new radio company called OmniAmerica Communications.

THE TURNING TIDE

By the early nineties, many new radio broadcast groups had difficulty meeting their debt service requirements. In some cases, banks took advantage of loan covenants—financial performance criteria on which loans are based—to demand full payment because of changing lending regulations by the federal government. In effect, the banks were calling in their loans. Many radio companies were also forced to restructure their deals either by seeking more equity or selling off parts of the company. For example, Jeff Smulyan, chairman of Emmis Broadcasting and owner of the Seattle Mariners baseball franchise, sold the team as well as some of Emmis's radio station properties—including its top biller, WFAN-AM New York—to help Emmis retire most of its senior debt. Emmis continues to own FM stations in major markets, including New York. (See Chapter 4 for more details on the sale of WFAN.)

Other entrepreneurial radio concerns, such as Westwood One and the Satellite Music Network, before it was acquired by Capital Cities/ABC, had already turned to the public market as an additional source for funding.

On Wall Street, investment houses started scrutinizing all potential broadcast transactions more closely, especially after the "Black Monday" stock market crash in October 1987. The broadcast financing market was also affected by the December 1989 downfall of the New York investment house Drexel Burnham Lambert, a major junk bond issuer.

With several media deals in the 1980s tied to so-called junk bonds (a form of subordinate lending consisting of short-term, high-yield/high-risk bonds), the decline of the junk bond market impacted new radio companies. Some financial and investment institutions, including major insurance companies, that dealt with both the radio and television industries entered a stage of *lending retreat*.

Faced with both a major slowdown in the economy and deficient capital due to loan overextensions in the 1980s (not just in media but also in commercial real estate), the major banks had become extremely cautious in their review of proposed radio deals.

Further complicating potential investments in the broadcasting industry was the temporary classification of most major deals as *highly leveraged transactions*, or *HLTs*, by the banks. Because the banking industry was having loan problems, this classification or new reporting standard—issued as a policy statement in 1990 by the Federal Reserve Board and other related federal agencies—gave regulators a way to measure the degree of risk or debt in a bank's lending portfolio. By their nature, many media deals fell into the HLT classification.

According to *Broadcasting & Cable* magazine, HLT loans were those made to fund a "buyout, acquisition or recapitalization" completed after January 1, 1987, if the transaction's "original financing package" was at least $20 million and met several asset and liability criteria.

If regulators thought it necessary, they could have restricted a bank's level of participation in HLT's. However, according to *Broadcasting & Cable*, broadcasters complained that the HLT definition "ignored cash flow on which the media industry has typically been valued." The results of the ruling on the radio industry were devastating.[10]

The observations of Richard J. MacDonald (a media analyst for Wasserstein Pirella Securities, a New York investment firm), who wrote about the business side of radio in the Freedom Forum Media Studies Center's 1993 *Journal*, put the situation in perspective:

> Credit simply vanished. Banks stopped lending and, to comply with rules, sought to reduce their highly leveraged loans as a percentage of total portfolios. Radio lending was singularly victimized in this process. As the bulk of outstanding HLT loans were enormous credits to companies like Time Warner, McCaw Cellular, and others, banks could not call in such loans without jeopardizing their entire portfolios. But radio credits, tiny in comparison, could be called in without any fiscal repercussions except, of course, to the radio industry.[11]

The HLT definition for banks was eventually dropped by the U.S. government.

Another aspect to this fiscal-tightening era involved the standard cash flow calculation for determining how much money banks or senior lenders would lend for a typical radio station acquisition. It fell from an average multiple of six to seven times a property's cash flow to an average of four to five times cash flow, causing downward pressure on station values. Cash flow calculations are the main method of determining a senior loan for a radio station purchase. (Refer to "Determining Value" in Chapter 2 for an explanation of various radio valuation factors.)

Even the financial terminology among lenders changed from lending against *leading* cash flow—projected cash flow for the next year, 12-month fiscal period, or for several years—to *trailing* cash flow—the cash flow recorded in the last calendar year or the previous 12-month fiscal period.

The residual effects of the HLT banking regulations, along with a sluggish economy, led to a substantial decrease in the number of lenders investing in broadcast properties, thereby lessening the amount of capital available to potential radio buyers, particularly first-time buyers.

Also complicating matters for lenders were that radio's real property assets were relatively limited in scope—the studio and transmitter, for example—compared to other businesses or industries. Most of a radio station's assets are intangibles—its FCC license and ability to attract listeners for advertisers. In addition, there is no security interest in a license, which reduces the amount of collateral available for loans. Major lenders or banks, therefore, cannot foreclose on an FCC license, only on a station's hard assets. A lender could also attempt to file for an involuntary receivership, allowing a third party to temporarily operate the station until it was resold.

As one would expect, financial industry executives favor attaching a security interest to broadcast licenses, saying it would create more opportunities for financial funding, particularly in a tight lending environment.

WHAT WENT WRONG: RADIO'S NADIR

Financial industry executives point to an overanticipation of retail revenue and market growth by many new radio station and group buyers as the cause of failed budget projections. Meanwhile, many senior lenders became speculators who were too eager to finance those broadcasting managers without a tenacious long-term marketing plan. Here is what prominent media broker Gary Stevens, of Gary Stevens & Co., said about such arrangements:

> In many cases, there were a number of executives on both the buying and lending side of the fence that didn't fully understand the dynamics of the radio business and the intense volatility (ratings variances) of a radio station. A television station, on the other hand, can be more financially dependable, especially if there is a network affiliation because its ratings are tied to the programs that the particular network distributes.[12]

Although annual radio advertising revenues continued to grow throughout the 1980s, the increments were not in line with the projections made in these radio deals several years earlier. Eventually it became difficult to routinely buy and sell radio properties. As previously noted, radio suffered the fallout of the HLT financing guidelines for banks, which were instituted primarily due to their poor real estate investments.

One could easily draw an economic comparison between the radio and real estate industries in the United States during this time. Both initially benefitted from a financially explosive decade, trading in what can best be described as an artificially inflated marketplace, followed by a financially stagnant period in an uncertain economy.

Broadcasting & Cable described the business world as it impacted the broadcasting industry at the beginning of the 1990s:

> By most accounts, 1990 was a year of transition. It was a year when the bull market of late 1980s speculative buying caught up with the real-world market realities of budget woes. . . . New definitions of government regulations on the banking industry greatly altered the dyspeptic buying patterns of the 1980s, while economic sluggishness driven by automotive and real estate downturns and retail softness yielded lower consumer spending than previously anticipated.

The residual effects of this financial crunch were felt at all levels of the radio industry, which had grown too fast too soon. Station trading dollars fell from a record-setting high of $3.41 billion in 1988 to $742 million in 1991.[13] And in 1992, for the first time in 30 years, radio advertising revenues were down from the previous year. (See Figure 3.2.)

ANNUAL RADIO REVENUE IN MILLIONS

Year	*Network*	*National Spot*	*Local/Retail*	*Total*
1993	$407.0	$1,629.0	$7,532.0	$9,568.0
1992	377.0	1,479.0	6,899.0	8,755.0
1991	440.0	1,573.0	6,578.0	8,591.0
1990	433.0	1,626.0	6,780.0	8,839.0
1989	427.0	1,530.0	6,463.0	8,420.0
1988	382.0	1,402.0	6,109.0	7,893.0
1987	371.0	1,315.0	5,605.0	7,292.0
1986	380.0	1,332.6	5,313.1	7,025.7
1985	328.7	1,319.4	4,915.0	6,563.1
1984	288.0	1,184.4	4,412.0	5,884.4
1983	253.5	1,022.8	3,739.0	5,015.3
1982	217.5	909.4	3,365.0	4,491.9
1981	195.9	854.3	3,007.0	4,057.2
1980	157.9	746.2	2,642.9	3,547.0
1979	138.5	637.3	2,396.6	3,172.4
1978	126.4	589.7	2,179.2	2,895.3
1977	118.1	521.3	1,873.1	2,512.5
1976	92.2	494.6	1,639.3	2,226.1
1975	72.7	416.3	1,403.3	1,892.3
1974	60.3	386.8	1,308.8	1,755.9
1973	59.4	382.3	1,205.4	1,647.1
1972	65.0	384.3	1,098.4	1,547.7
1971	55.1	378.0	954.6	1,387.7
1970	48.8	355.3	852.7	1,256.8
1969	50.9	349.6	799.9	1,200.4
1968	54.7	342.2	733.4	1,130.4
1967	58.2	298.3	641.2	997.6
1966	57.4	292.6	607.6	957.7
1965	54.3	261.3	553.0	868.7
1964	54.0	244.1	504.2	802.3
1963	51.5	231.0	465.0	747.6
1962	44.9	218.2	434.2	697.3
1961	47.7	205.6	397.7	651.0
1960	44.9	208.0	401.6	654.5

Source: FCC through 1980/RAB from 1981.

Figure 3.2

Broadcasters also complained that the FCC had granted too many new channel assignments for new stations during the 1980s, particularly on the popular FM band. Through what is notoriously referred to in the industry as the *Docket 80-90* (FM station assignment) ruling, many of the smaller market radio outlets became overcrowded. (See Chapter 4.)

The inaction by financial institutions that brought radio station trading and the industry itself to a near standstill was due to three key factors:

- *A credit crunch resulting mainly from overextended real estate loans, which affected all industries,*
- *The classification of most media deals as HLTs by banks,* the definition for which has since been discontinued by the Federal Reserve Board, and
- *Substantially reduced advertising spending, which jeopardized many radio station deals completed during the free-spending eighties.*

> I don't think the radio industry ever faced anything like this. You had massive debt, a flat advertising market, a lack of buyers and increased competition. These are fundamental changes that made it very hard to do business.—Ed Christian, president of Saga Communications[14]

According to Christian, however, the early nineties was a good time for owners to buy quality radio facilities with receding cash flows in distressed markets to position themselves in key regions when the economy turned upward. This philosophy was part of Saga's ownership strategy as it acquired stations in recession-hit New England communities.

The rationale for such an approach is that the radio station business already has a proven revenue growth record. With the proper positioning, therefore, there is the potential for a quick *turnaround* for cash-poor radio properties.

Through radio's turbulent period, the industry was not without its major entrepreneurial success stories. In one example Infinity Broadcasting, under the watchful eye of president and CEO Mel

Karmazin, steadily built a portfolio of successful radio stations in many of the country's larger cities, eventually operating stations in each of the top 10 markets. Infinity established a strong major market presence when it purchased WJIT-AM and WKTU-FM New York and WYSP-FM Philadelphia in the early eighties. (WKTU is now WXRK.)

As the industry progressed through the eighties into the nineties, Infinity, whose only business is radio, pursued strong cash flow–driven stations in key markets, especially where it already was a significant player. By industry standards, the company paid premium prices for these properties. Karmazin has frequently compared this bullish approach to desirable station acquisitions to the purchase of oceanfront property. The company's blue-chip ownership strategy and high earnings performance has repeatedly impressed Wall Street analysts and financiers alike.

Another broadcasting company that withstood the test of the eighties was Clear Channel Communications, a radio and television concern based in San Antonio, Texas. During this decade, the company gradually assembled one of the industry's more profitable medium-market radio group operations. Established in 1972 by former investment banker Lowry Mays, who now serves as president, Clear Channel Communications purchased stations, primarily across America's Sun Belt region, based upon their long-term cash flow potential. The company adopted a conservative acquisition posture by shying away from the highly leveraged deals that were dominating the industry. It took a "wait and see" attitude when station values soared. In so doing, the company avoided the debt service problems that eventually engulfed many radio owners, positioning itself for growth in the nineties.

By 1992, an economic rebound to the third U.S. recession since 1980 was announced by the federal government. Simultaneously, the FCC took a major step to realign the U.S. radio industry with a complete overhaul of all radio ownership rules. The move paved the way for a greater concentration of local radio ownership by allowing multiple, same-market station ownership on the same band while vastly increasing national ownership limitations. This gave radio broadcasters more flexibility in competing with other media for advertising. (See Chapter 5.)

The radio owners who prudently operated their businesses during the tumultuous late eighties and early nineties were about to regain control of their industry.

Before discussing radio's newest structure, however, it is important to first address AM radio's ubiquitous influence in shaping modern-day programming.

NOTES

1. Peter F. Drucker, *Innovation and Entrepreneurship; Practices and Principles* (New York: Harper and Row, 1985).
2. Based on an interview with Frank Osborn of Osborn Communications, New York.
3. "Metromedia Sold in LBO," *Broadcasting Magazine*, March 31, 1986, pp. 33–34.
4. Ibid.
5. "Robert F. X. Sillerman: Radio's Dealmaker," *The Pulse of Radio* (now *Radio Ink*), Jan. 15, 1990, p. 22.
6. "Beginning a Legacy," *Broadcasting Magazine*, Nov. 18, 1985, p. 44.
7. "Metropolitan Slims Down to Service Debt," *Broadcasting Magazine*, Feb. 29, 1988, p. 63; "$300 Million Sale Would Set Radio Only Record," *Broadcasting Magazine*, April 11, 1988, p. 36.
8. "Sillerman's Radio Reorganization," *Broadcasting Magazine*, July 4, 1988, pp. 39–40.
9. "Record Breaking Deal Set in Motion," *Broadcasting Magazine*, April 24, 1988, p. 28; "Radio's Newest Force: Group W Sets Its Sights on the '90s," *Broadcasting Magazine*, Dec. 11, 1989, pp. 35–38.
10. "HLT: A Three-Letter Worry," *Broadcasting Magazine*, Feb. 4, 1990, p. 44.
11. Richard J. MacDonald, "On the Business Side—An End to Radio Romance," as published in the Freedom Forum Media Studies Journal: *Radio: The Forgotten Medium*, p. 163. (Richard J. MacDonald is a media analyst for the New York investment firm, Wasserstein Perella Securities.)
12. Based on an interview with Gary Stevens of Gary Stevens & Co., New Canaan, Conn.
13. Based on station trading data from Paul Kagan Associates.
14. Based on an interview with Ed Christian of Saga Communications, Grosse Pointe Farms, Mich.

4

The AM Dilemma

LOSING THE EDGE

Long before FM radio was established, there was AM. However, the medium that first transmitted live entertainment and news reports from around the world faced a major crisis in the 1980s: the erosion of listening and, consequently, advertising support.

During the halcyon days of rock 'n roll in the 1950s and 1960s, AM radio was king. High-powered AM signal transmissions, which travel through the sky reflecting off the ionosphere, cover a wider area than FM signals, which travel close to the ground, or line of sight. During this time, most homes and nearly all car receivers were only equipped to receive AM signals.

With small audiences, the handful of FM stations that existed, primarily co-owned with AM powerhouses in the same market, either simulcast the AM format or offered programming that was not a large ratings winner, like classical music.

As the number of new FM stations continued to expand, however, in 1964 the FCC decreed that FM stations commonly owned with AM stations in the same market should have at least half of their programming differentiated from sister AM stations to create greater diversity in the listening marketplace.

As a result, FM radio stations, many of which were being transmitted in stereo, became a major force in radio. These stations became bastions of experimental music programming, especially music formats that featured new forms of rock music. They featured longer, uninterrupted music segments and more new artists and selections than the average AM contemporary (hit) music station.

By the early 1970s, the young adult post–World War II baby boomers (people born between 1946 and 1960) were tuning to FM stereo music stations in droves. Both radio and television broadcast advertisers earnestly targeted this demographic group.

Then in 1979, what had been unthinkable just a few years earlier occurred: FM's audience share, at 52 percent of the total listeners in the United States among people over 12 years of age, surpassed AM's audience.

Modern-day national rating trends for AM and FM listening audiences have been compiled since 1972 through the RADAR audience survey report conducted by Statistical Research Inc. At that time, AM radio commanded a substantial 75 percent of total radio listening, compared to FM's 25 percent share. In 1973, FM listening increased to 28 percent of the total. But by the late 1980s, FM listening levels already hovered near 75 percent. FM's share of radio listening finally leveled off at about 77 percent in the early 1990s.[1] (See Figure 4.1.)

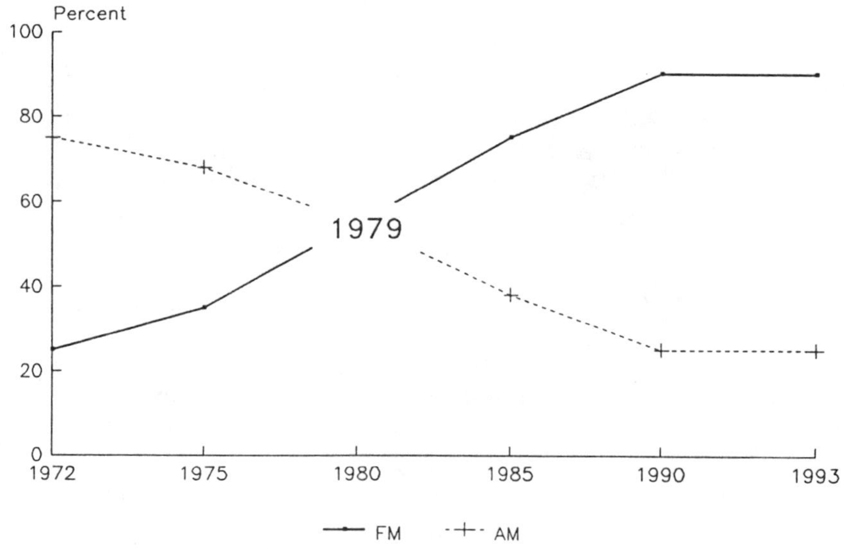

FIGURE 4.1. AM versus FM listening

THE DRAMATIC RISE OF COMMERCIAL FM VERSUS AM STATIONS SINCE 1970

Year End	*AM Stations*	*FM Stations*	*Total Stations*
1990	4,987	4,392	9,379
1985	4,718	3,875	8,593
1980	4,589	3,282	7,871
1970	4,323	2,196	6,519

Source: FCC/RAB.

FIGURE 4.2

The main reason for this increase in FM audience share stems from the U.S. government's move to supply all sections of the country with a variety of radio programming sources by taking advantage of the fledgling FM stereo band. From 1970 through 1980 the number of new commercial FM stations on the air exceeded the number of new commercial AM stations by more than 800. (See Figure 4.2.)

The increase in new FM stations also translated into considerable gains in advertising revenue for FM radio, which, based upon FCC data, rose from $85 million to $1.216 billion during the seventies.

Those numbers continued to increase in the 1980s due to an FCC ruling—known as *Docket 80-90*—that boosted the availability of FM broadcast assignments. Issued in 1984, the FCC's objective with this ruling was "to meet the demand for new FM stations, particularly in those communities lacking aural service." AM listening and revenue levels continued to tumble. This FCC ruling became the precipice for the impending sagging fortunes of AM radio.

(The FCC discontinued the required accounting for AM and FM radio revenue estimates in 1980. There has since been several industry-wide voluntary attempts to organize the collection of AM and FM radio billings, but these efforts subsequently failed. By the best estimates, FM radio revenues in the early 1990s amounted to between 60 percent and 70 percent of all radio revenue.)

By the late 1980s, radio station groups channeled most, if not all, of their resources into FM radio. During this time some AM stations that were commonly owned with FM stations in the same market were sold to different radio groups while others were marketed to advertisers along with their sister FM properties. These AM–FM stations are known as *combos*. Co-marketing sister AM and FM stations, at times referred to as *combo selling*, gave the owners of foundering AM facilities an edge because they could combine AM and FM audiences for ratings-conscious buyers for little or marginal cost. Separate AM–FM rates were also offered. (It became more difficult to discount rates in radio's new, expanded duopoly era [as discussed in Chapter 5] because combos began to take the form of stronger AM–AM and FM–FM properties.)

Eventually AM-only radio stations, which are sometimes called *stand-alone properties* (not part of co-located FM station deals), were no longer desirable to operate. There were the exceptions, however, like long-established major market AM outlets such as KDKA in Pittsburgh.

The declining financial stock of AM radio was underscored in a 1989 station trading study conducted by the New York City–based investment concern Hoffman/Schultz Media Capital, and cited by the National Association of Broadcasters at an FCC AM improvement hearing. The study concluded that the average price for AM radio stations was dropping at an annual rate of 4 percent, while the value of FM stations was rising 20 percent a year.[2]

With soft advertising sales stemming from the economic recession of the late eighties and early nineties, the profitability of many AM radio stations was quickly deteriorating, particularly those stations in smaller markets. The economic woes of AM radio surfaced in a 1991 survey of radio station finances, jointly conducted by the National Association of Broadcasters and the Broadcast Cable Financial Management Association (BCFM).

The survey showed pretax profit (or net revenues less expenses) for AM stations in the United States was down dramatically in 1990, compared to radio's financial frenzy period during the 1980s.

For example, the analysis shows that for daytime AM stations—those stations that only have authority to operate during daytime hours in order to reduce nighttime signal interference with full-time AM stations—pretax profits fell from an average of $7,969 in 1987, to $1,418 in 1988, and to an average of –$13,865 in

1990. Pretax profits for full-time AM outlets went from an average of $53,939 in 1987, up to $90,957 in 1988, and down to $29,175 in 1990.[3]

Financing for potential buyers of AM-only radio stations, especially for those outlets in the smaller markets, became almost nonexistent as many of those properties were now generating negative cash flow.

Some veteran broadcasters, however, refused to walk away from a good fight. As music programming swung to FM during the 1980s, AM stations found their niche in informational programming, which generally attracts an over-35 demographic age group.

New niche or specialized nonmusic programming concepts had been developed for an information-seeking audience. Some have succeeded while others failed. Among these unique programming ventures were all-weather, all-business, all-sports, and motivational radio formats.

Enter into the radio broadcasting lexicon the term *fragmentation*. An AM station plagued by low ratings could now market a new, specialized format to the advertising community as a qualitative concept sell. Instead of buying mass audience numbers or broad-based demographics, advertisers and agencies could buy a highly-defined programming concept for targeting a precise audience group. This selling technique kept many AM stations afloat in an extremely competitive media marketplace amidst a changeable economy.

ALONG THE ROAD TO SPECIALIZATION

AM radio is no stranger to program specialization. To begin with, it was the birthplace of formatic radio known as Top 40.

Although FM stations mainly aired classical music in the late 1950s, the first significant move toward the modern-day concept of specialization or highly defined programming occurred on the AM band in 1961 when the all-news format was launched by Gordon McLendon, one of the country's more innovative radio programmers, over XETRA-AM Tijuana, Mexico. While based near the U.S.-Mexico border, XETRA's signal was aimed at nearby Southern California, covering San Diego to Los Angeles.

McLendon's all-news format is best characterized as a "rip 'n' read" or headline news service with announcers reading copy from the various news wire services.[4] The station attracted a large cumulative audience even though listeners only tuned in for short periods during the day. The McLendon format lasted for several years. It was the first time a radio station focused entirely on a programming element other than music or religion; news historically had been no more than a 10- to 15-minute hourly broadcast. McLendon later aired the all-news format in Chicago.

New York, however, would soon become the mecca for what has turned out to be the modern version of all-news radio, featuring street reporters and more in-depth reporting. This version of the all-news format was first implemented by Westinghouse Broadcasting over WINS-AM New York in April 1965. WINS had been a Top 40 rock music station, but it was losing audience to its local competitors, WABC-AM and WMCA-AM, forcing Westinghouse executives to search for a new format.

Influenced by McLendon's success, Westinghouse switched WINS to an around-the-clock news format. The move proved to be the right one that November when a major electrical blackout occurred in the Northeast. Most residents in the New York metropolitan area with portable radio sets tuned to WINS for its blackout coverage. All-news radio had arrived in the nation's largest city. Westinghouse also started an all-news format over company-owned KYW-AM Philadelphia in 1965.

Two years later CBS launched an all-news format on its flagship New York station, WCBS-AM, that varied from Westinghouse's design. CBS relied on the resources of the CBS Radio Network to present an all-news format with a national slant. Westinghouse, which did not own a network and had no network affiliations for WINS and KYW, offered a more local all-news format. Both programming forms soon thrived, and New York became the first market to support two all-news radio stations. Other Westinghouse and CBS radio stations soon followed with all-news formats. Both the Westinghouse and CBS formats have maintained their distinct presentations over the years, but Westinghouse stations have since signed network affiliations.

The all-news format was originally dismissed by some broadcasting executives as too narrow for audience appeal, even for a specialized medium like radio. What they failed to anticipate was that as the world got more complex during the turbulent sixties, the American public would develop an insatiable quest for current events.

With most music formats shifting to FM bands, all-news became a programming staple of AM radio. Because it was considerably more expensive to produce than other formats, big-budget AM radio stations in major cities were more inclined to program all-news first. By 1990 almost every major city in the United States had at least one all-news or news/information AM station on the air.

THE TALK EXPLOSION

All-news was not the only nonmusic format being assimilated into the American radio culture. Telephone talk or talk radio evolved as an alternative AM radio format. Although talk radio did not reach its stride among listeners until the late 1970s, the format had been airing for some time. In Southern California, talk radio flourished. KABC-AM Los Angeles put a talk format on the air in 1960 and the station became one of the more listened to radio outlets in Los Angeles.

In 1978, Mutual Broadcasting decided to bring the local flavor of the telephone talk format to a national audience. The company tapped Larry King, a talk radio personality from WIOD-AM Miami, to host an all-night talk show with both celebrity guests and audience call-in segments over the Mutual Radio Network. King was an overnight success, and talk radio became a national phenomenon. The show started with a network of 28 stations and by the early 1990s was distributed to more than 300 stations.

The Larry King Show eventually switched from overnight to daytime (3:00 to 6:00 P.M. Eastern time), and in 1994 King left the daily routine of hosting a live network radio program, to concentrate more on his other media duties, including his role as CNN talk-show host. Comedian David Brenner took over King's afternoon time slot. (Westwood One, of which Mutual Broadcasting is a part, began offering the audio portion of CNN's *Larry King Live* evening broadcast in June 1994.)

If one event in radio signaled the end of contemporary music's reign on the AM band, it was the format switch of WABC-AM New York from Top 40 to talk in May 1982. "There is an unfulfilled appetite for additional talk radio programming in the New York market," then WABC vice president and general manager Al Racco was quoted as saying.[5] The move by WABC to talk is also believed to have been influenced by then ABC Radio president Ben Hoberman, who arrived several years earlier from KABC Los Angeles.

WABC was the premiere rock 'n roll music station in the United States, particularly throughout the 1960s when it registered enormous teenage and adult audience ratings (through roughly age 49) with its tight music playlist of pop, rock, and Motown hits, replete with lively jingles and creative on-air promotions.

What hastened the mass proliferation of music stations across the FM dial was a new form of dance music called disco, which became an instant FM format sensation in 1978 after the release of the movie *Saturday Night Fever*. Disco FM stations began chipping away at AM radio's contemporary music audiences, including WABC's.

Besides Mutual, talk programming for AM radio became the focus of two other major networks—ABC and NBC. During the early 1980s, NBC launched a satellite talk programming service called Talknet, and ABC established a similar service called Talkradio. Talknet was designed for nighttime airing; while Talkradio was principally designed for daytime audiences.

In 1990 ABC Talkradio repositioned itself as a nighttime/weekend service with leading talk personalities Tom Snyder and Sally Jesse Raphael competing directly with Talknet. The ABC service was subsequently scaled back to weekend programming, however, without the high-profile personalities of Snyder and Raphael. Meanwhile, a multitude of other national talk services have been launched.

Network executives at both ABC and NBC believed that as the 1980s progressed, most AM radio stations would be looking for quality talk programming. That belief held true as both networks were able to successfully coexist in the radio program marketplace. AM broadcasters discovered that by addressing issues with an immediate or direct impact on the typical listener, such as personal, financial, or legal advice, a two-way telephone talk format could establish a respectable ratings niche. Local and national talk radio became the sounding board for the views of its hosts and its listeners. It even became a powerful tool for influencing congressional legislation.

The popularity of two-way telephone talk programming continued to soar as the medium entered the 1990s with the emphatic ratings success of conservative talk show host Rush Limbaugh. Limbaugh, who does not shy from sharing his political (often considered politically incorrect) views with listeners, originates a daily program from the studios of WABC-AM New York that is distributed, via satellite, by EFM Media to over 600 radio stations. (EFM Media was founded by Edward F. McLaughlin, former president of the ABC Radio Networks.) What is significant about Limbaugh's syndicated broadcast is that it airs during the afternoon (noon to 3 P.M. eastern standard time), which, for the most part, had proven to be local talk radio's domain.

> For all we hear today about Rush Limbaugh's phenomenal success, few remember that it does have its antecedents. He has revived—perhaps perfected—the art of using radio to connect, an art most in the business seemingly have forgotten.—Tom Lewis[6]

A direct result of this achievement is that more listeners are tuning to AM talk radio, prompting more AM stations to turn to the talk format.

On the heels of Limbaugh's success in the early 1990s came the national syndication of another controversial talk show program, Infinity's G. Gordon Liddy broadcast, designed for daytime airplay. The show originates from WJFK-FM Washington, D.C. CBS Radio also entered the talk radio market for the first time during this "talk explosion" period with its launch of *The Gil Gross Show*, a three-hour evening broadcast.

Program fragmentation on radio had been refined even further with the introduction of the all-sports format in New York City. Seizing upon the informational trend in AM radio, Emmis Broadcasting introduced a sports talk and play-by-play format in 1987 on one of its new radio properties, WHN-AM. The call letters were changed to WFAN-AM, which allowed the station to promote itself as "The Fan."

The New York market seemed like the perfect setting to undertake such a specialized programming endeavor because the city supported at least two major team organizations in baseball, football, basketball, and ice hockey. The station soon established a strong core of sports listeners, becoming a radio franchise in New York.

After Emmis purchased the NBC radio station group from GE, which included WNBC-AM, the company shifted the WFAN dial position to the one held by WNBC (660 kHz) because of its clear channel status. (Clear channel frequencies serve large geographic areas with up to 50,000 watts of power and are protected from interference within their primary service areas and, depending on their government classification, secondary service areas.) The WNBC call letters, one of the oldest in the country, along with its contemporary music format, were dropped for WFAN and the all-sports format. The Emmis station switch also marked the demise of the last contemporary music station on the AM dial in New York.

After two years of operating in the red, WFAN turned the profit corner in the early 1990s, becoming one of the top five revenue producing radio stations in the United States. In late 1991, WFAN and its all-sports programming franchise was purchased by Infinity Broadcasting for $70 million, a record price for a stand-alone AM radio station.[7]

The sports format did not end in New York, however. An increasing number of AM radio stations across the United States emphasized sports talk and play-by-play programming in their formats. In addition, an ESPN radio sports network was launched by ABC—a major owner of the ESPN cable networks—to serve the increasing interest in sports programming.

While music formats became indigenous to FM radio, a smattering of rock 'n roll oldies and country formats, two types of music usually associated with the maturing baby-boomer population, are still heard on AM. In fact, oldies was one of the fastest growing U.S. radio formats in the late 1980s.

The oldies format, consisting of familiar Top 40 hit songs from the fifties, sixties, and early seventies, succeeds in reaching most of the 25- to 54-year-old audience, which, from an overall buying perspective, has been the most sought after demographic group among broadcast advertisers. The same is true of many mainstream country radio formats. (A shift toward target marketing within the entire audience spectrum is becoming more evident among advertisers.)

Another music format that found a home on AM during the eighties was the big band or nostalgia format, as it is referred to by industry programmers. This covers popular music primarily from the late 1930s through the 1950s.

CONTINUITY

Although AM listening levels declined over the years, longstanding major market AM radio stations that accent community programming, otherwise known as "full-service" stations, have been able to maintain a loyal listener base from generation to generation. Their longevity is attributed to the retention of both call letters and basic programming fare over a sustained period.

Some of these highly recognized AM radio outlets include KMOX-AM St. Louis, WOR-AM New York, WGN-AM Chicago, WJR-AM Detroit, WMAL-AM Washington, WCCO-AM Minneapolis-St. Paul, WLW-AM Cincinnati, WBZ-AM Boston, and KDKA-AM Pittsburgh—the last two being among the first radio stations in the United States with commercial broadcast licenses. Many major market AM information-oriented outlets have remained among the top revenue billing radio stations in the country. According to an analysis by Broadcast Investment Analysts (BIA), a Chantilly, VA-based financial services consulting firm for the media industry, six of the top 10 billing stations in 1993 were AM stations.[8]

There were exceptions to this continuity trend, however. In Philadelphia, for example, CBS-owned WCAU-AM (an institution for years in that city) dropped an established informational format in 1990 to simulcast the oldies format of WOGL, its sister FM station. WCAU also adopted WOGL's call letters. At the time, WCAU's format switch underscored the changing economics of the radio industry and the increasing costs associated with operating sister AM and FM radio facilities in large cities. By 1994, the station had switched to an all-sports format.

NEW PARTNERSHIPS

Simulcasting is no longer limited to commonly owned AM and FM stations. Beginning in 1990, radio station operators signed program service and/or marketing agreements, known in the industry as local marketing agreements, or *LMAs* for short, with nearby stations, thereby giving new meaning to the term *broadcast affiliate*.

These new affiliates can either be located in the same market or the same region and on the same service band—that is, two AM or two FM stations. The owners of these affiliates have legal and financial control over their properties and they maintain their individual station identities.

An LMA is a time-brokered agreement between stations that can take the form of programming time and/or advertising time. Basically, the originating or principal station in an LMA agreement pays the affiliate a monthly fee to either partially simulcast programming or air original/satellite programming. This is different from a satellite format network affiliation arrangement where an affiliate pays for programming. The originating station through the LMA deal can strike an arrangement with the leased station for either handling or sharing advertising sales.

The benefits of an LMA to the originating station are an expanded coverage area and the potential for increased sales opportunities. Many struggling AM radio operators have been the recipients of these deals, reestablishing a steadier flow of income.

LMAs have been sanctioned by the FCC. In markets where these arrangements take place, the barriers to competition are breaking down. These affiliation deals allow radio stations to focus their marketing efforts on local media competition other than radio. (A further discussion of LMAs appears in Chapter 5.)

STEREOPHONIC AM

As the AM band went through its metamorphosis as radio's secondary service, the push for stereophonic broadcasting was well under way. AM executives believed stereo would give them the competitive edge necessary to achieve parity with FM radio's high acceptance from the American public.

Initially, five electronics manufacturers, Belar, Harris Corp., Kahn Communications, Magnavox, and Motorola, developed AM stereo systems. In early 1980 the FCC approved AM stereo and chose the Magnavox system as the standard. But two years later, under the Reagan administration's deregulation push, the FCC backed away from endorsing one AM stereo system, deciding instead to let marketplace forces dictate the choice.

This move differed greatly from the FCC's authorization of one FM stereo standard, a three-dimensional Zenith system, in 1961 when the fledgling FM medium was still in its infancy. Ironically, at the time stereo was viewed as a necessary step to give FM a competitive position with AM.

AM stereo could have been as old as FM stereo. The same year stereophonic transmissions of FM signals were approved, the FCC turned down a petition from Philco, RCA, and Kahn to establish AM stereo broadcasting. "There is little evidence of public need or industry desire . . . with respect to stereophonic transmissions by standard (AM) broadcast stations," the FCC said.

The ramifications of the FCC's AM stereo marketplace decision were felt throughout the 1980s as competing AM stereo equipment manufacturers fought for acceptance among radio broadcasters. This infighting forced most stereo receiver manufacturers to indefinitely halt plans to produce AM stereo sets.

By the beginning of the 1990s, two AM stereo systems manufactured by Kahn and Motorola were still in operation. Due to a 1992 congressional ruling, however, which among other things, required the FCC to name an AM stereo standard, the FCC finally endorsed the Motorola C-QUAM system,[9] also the standard in Japan, Canada, and Australia.

By this stage, it was evident that AM stereo broadcasting was not the panacea its operators had envisioned for bolstering the popularity of the band. While many AM stereo stations promoted themselves on the air as broadcasting in stereo, the impact on the general listening audience was negligible. AM stereo transmissions were not widely discernible by radio listeners due to a lack of AM stereo receivers in the marketplace. Exacerbating the situation was the fact that most AM outlets had switched to a nonmusic format, which does not particularly lend itself to the advantages of stereo.

Although it is hard to draw an exact figure for any given point in time, radio and receiver industry executives agree that only a small percentage of the approximate 5,000 or more U.S. commercial AM stations are transmitting in stereo.

REGULATORY RELIEF

Many other technological changes designed to improve the fidelity of AM radio signals by reducing adjacent-channel and/or noise interference, a continual problem for many AM stations, have been implemented by the FCC as part of an overall AM rescue plan for the 1990s. To relieve AM station signal congestion, the International Telecommunications Union (ITU), an international regulatory body for communications issues, unveiled plans to extend the AM band in the Western Hemisphere from 1605 kHz to 1705 kHz, opening new frequency slots for 250 to 300 AM stations in the United States.

The FCC is giving preference for these new channels to AM daytime-only stations licensed in communities of 100,000 or more and to those that are now suffering from adjacent-channel interference. To ensure a smooth transition to their new dial positions, migrating stations will be permitted to simulcast on both the old and the new channels for five years.

Meanwhile, advanced AM radio receivers that have a wide range of new fidelity features, including stereo and expanded band reception, are being developed. Under guidelines set by the National Radio Standards Committee (NRSC), a joint committee of the National Association of Broadcasters and the Electronic Industries Association, new high-performance AM radio receivers that met NRSC recommended specifications can carry a special logo: *AMAX*— an acronym for "maximum AM." According to NAB, this logo signifies that a receiver provides the "very best AM signal quality attainable from a state-of-the-art receiver"—a quality comparable to FM signals.

Among the receiver manufacturers producing advanced AM radio sets are Denon America; General Electric, producing a series of new portable AM receivers marketed under the name of "Superadio"; and Sony, which is manufacturing a personal Walkman radio certified with the AMAX logo.[10]

AM HORIZON

Besides technological and regulatory improvements, some radio broadcasting executives see AM radio poised for a comeback for a variety of reasons:

- An aging of the U.S. population, specifically the baby-boomer age group,
- The decline of daily newspapers, making AM radio a stronger information medium,
- The enormous geographical reach of AM signals that benefits brand name advertisers,
- The stabilization of AM audience listening share, which according to Statistical Research's RADAR reports, held steady at 23 percent in the early 1990s,
- The surging popularity of talk radio hosts such as Rush Limbaugh,
- The high values being attached to select major market AM stations, evident by Infinity's 1991 purchase of WFAN-AM New York.

NOTES

1. FM's share of radio listening in the early 1990s varied from 76 percent to 78 percent.
2. "New FCC Looks at Broadcasting's Oldest Medium," *Broadcasting Magazine*, Nov. 21, 1989, pp. 31–32.
3. Data from the National Association of Broadcasters-Broadcast Cable Financial Management Association 1992 Radio Financial Survey.
4. Philip Kiersted, *All-News Radio*, (Tab Books, 1980), 20.
5. "WABC-FM New York Plans Switch to All-Talk in Early May," *RadioNews*, March 1, 1982, p. 1. (*RadioNews* was acquired by *Radio Business Report* in the mid-1980s.)
6. Tom Lewis, "Triumph of the Idol—Rush Limbaugh and a Hot Medium," as published in the Freedom Forum Media Studies' 1993 Summer Journal: *Radio: The Forgotten Medium*, p. 55. Tom Lewis is a professor of English at Skidmore College in Saratoga Springs, New York, and the author of *Empire of the Air: The Men Who Made Radio.*
7. "Radio Record: $70 Million for WFAN," *Broadcasting Magazine*, Dec. 16, 1991, p. 4; "Infinity Goes Public: Initial Offering Would Help Finance $70 Million Purchase of WFAN-AM," *Broadcasting Magazine*, Dec. 16, 1991, p. 18.
8. *The BIA Monitor*, Broadcast Industry Analysts, Jan. 10, 1994, p. 1.
9. "AM Stereo Rears Its Divided Head," *Broadcasting Magazine*, April 12, 1993, pp. 61–62; "FCC Says C-QUAM Is AM Stereo," *Broadcasting Magazine*, Nov. 1, 1993, p. 14.
10. "High Quality AM on a Budget," *Radio World*, March 24, 1994, pp. 1+.

Part II

5

Realignment of the Industry

THE 1990S: UNPROFITABILITY YIELDS NEW BLUEPRINT FOR OWNERSHIP

> The economies of scale dictate that duopolies are going to be a huge part of the business, if for no other reason than economically—it's what's selling on Wall Street. —Steve Goldstein.[1]

From a pure economic standpoint, the radio industry in the early 1990s was realigned into a three-pronged paradigm incorporating both management and programming operations. Radio's new structure evolved into

- Financially stronger companies through government expansion of station ownership, primarily leading to competitive market positioning with other local media as well as a reprisal in active trading and group mergers;
- Business alliances among former competitors;
- Programming of syndicated, locally produced shows—delivered primarily by satellite—featuring nationally known or highly rated personalities.

With the decline in radio profits (especially for AM stations) during the early 1990s, the FCC was petitioned by broadcasters and in 1992 expanded local and national station ownership rules for radio broadcasters. What hastened this action was a 1992 radio revenue survey, jointly conducted by the National Association of Broadcasters and the Broadcast-Cable Financial Management Association, which showed that in 1991 nearly 60 percent of all U.S. commercial radio stations were unprofitable, registering a decline in pretax profits.[2]

Many of these stations were located in the smaller sized markets, where factory layoffs and retail closings occurred during the country's latest economic downturn. The study proved that radio could be vulnerable to quick marketplace turns.

While other segments of the economy started a slow recovery, the survey clearly illustrated that the radio industry was mired in steep financial trouble, with many beleaguered operators plagued by sluggish advertising levels and soaring debt payments. The study was proof that the industry suffered from structural weaknesses and would benefit from relaxed ownership rules.

> This can be a rough and tumble business. It's a very competitive business. If you get a good, well-capitalized competitor [in your market], you may effectively blunt their attack, but your cash flow will be going down for a while.—Frank Osborn[3]

In 1992, the National Association of Broadcasters entered comments before the Securities and Exchange Commission (SEC) in the Matter of Small Business Initiatives, illustrating the disappearance of capital from the broadcast marketplace. (These comments also appeared in the statement of FCC Commissioner Ervin S. Duggan, who voted for expanded ownership rules.)

The association noted that half of the 115 senior commercial banks that were "actively engaged in broadcast lending" in 1989 no longer were making broadcast loans. They stated further that $2.2 billion in new broadcast financing had "shriveled" to $191 million by 1992.[4]

Both the radio broadcasting industry and government alike thought the FCC's move to expand local AM–FM station ownership restrictions (by broadening its so-called duopoly rule) and, simultaneously, extend national ownership limits would improve the competitiveness of the radio business with itself and other media. The original *duopoly rule* states that an entity can own only one AM and FM station in a single market. The rule had been enforced by the FCC as part of an overall government policy to dissuade media monopolies.

Radio station owners staunchly argued that their medium must now be able to compete more effectively for advertising dollars, especially with the emergence of cable and new media companies. They reasoned that the FCC's duopoly ruling would lead to more financially viable radio station companies through ownership consolidation. It would not mean a reduction in the number of on-air radio stations, the owners said. Rather, it would provide greater operational synergies for radio groups and, therefore, more diverse programming products—the latter issue a concern of Congress (see below).

The logic here is that a new investor-friendly climate for radio would be created through a concentration of successful ownership. The profitability of stations would again rise, leading to more quality programming for all radio audiences. These potential benefits to duopoly would be particularly salient given the by-products of the financial crunch of the early nineties—tighter cash flow and inventory controls. (See Chapter 6 for details on inventory control management.)

> The new rules will allow the industry to be more competitive. It's great to say that we need thousands of radio stations. . . . But if 60 percent of them can't survive, we haven't really fostered diversity; we have fostered lack of diversity.—Jeff Smulyan[5]

After analyzing the radio marketplace to evaluate the old national and local radio ownership rules, the FCC concluded that there indeed was "intense competition within the communications as a whole, and within the radio industry in particular."

The FCC went through two AM-FM ownership machinations before settling on its final blueprint for local and national limits. The FCC's first modification of radio station ownership was the 30-30 rule, which would have allowed radio broadcasters to own up to thirty AM and thirty FM stations nationwide (sixty stations total). The previous national limit had been twelve AM and twelve FM stations, or fourteen of each if the business was minority-controlled. This initial FCC ruling would have also permitted an entity to own an increased number of stations within a local radio market, depending on market size and the combined audience shares of stations.

Issued in March 1992, the ruling immediately met with resistance from several key congressional members including Rep. John Dingell (D-Mich.), then Chairman of the House Commerce Committee, which oversees telecommunications issues, and Sen. Ernest Hollings (D-S.C.), then chairman of the Senate Commerce Committee, who were concerned about the effects of ownership consolidation on program diversity, particularly in smaller markets. That summer, many members of Congress threatened legislation to roll back radio ownership limits to the 12-12 status quo if the FCC didn't act accordingly.[6]

Even NAB expressed concern about the increased station ownership allowances, suggesting a more conservative approach to safeguard competition. Under pressure, the FCC issued a revised and final version of the ruling in August 1992. Despite its judgment that its initial ruling was "fully justified," the FCC significantly scaled back the national ownership limits to eighteen AM and eighteen FM stations, which would then increase to twenty AM and twenty FM stations in September 1994. The FCC also said it would permit "attributable but non-controlling interests" in three additional AM and FM stations if those stations are controlled by small businesses or are more than 50 percent owned by one or more members of a minority group. "This offers a fresh opportunity to strengthen broadcast diversity and to attract capital to stations that badly need it," said FCC commissioner Erwin Duggan at the time the order was issued. (In early 1994, Erwin Duggan became the president of the Public Broadcasting Service.)

On the local front, the FCC provided specific guidelines for increasing station ownership, allowing up to two AM and two FM stations in markets with fifteen or more stations—down from three AM and three FM stations in its March ruling. Applicants would have to demonstrate to the FCC that any acquisitions in this kind of expanded duopoly deal would not exceed a 25 percent audience market share, as evidenced by ratings and other source data.

The 25 percent audience market criteria would only apply at the time of acquisition, however. So if the new combination of commonly owned stations' market share exceeds 25 percent later, divestiture generally would not be required.

In markets with fewer than fifteen stations, ownership of up to three stations is allowed, only two of which can be on the same band, provided that this group of stations represents less than 50 percent of the total number of stations in the market—the same as the March ruling. Only full-power commercial stations are counted toward the total number of stations allowed to be owned in a single market.[7]

These FCC rules went into effect on September 18, 1992. (See Figure 5.1 for a comparison of the old and new radio station ownership rules.)

As part of this ruling, one method used by the FCC to ensure programming diversity in radio is to limit simulcasting on commonly

NATIONAL LIMITS

Old Rules	*New Rules*
Overall limit of 12 AM and 12 FM stations (14 each if owned by minority-controlled group	Expanded first to 18 AM and 18 FM, then to 20 AM and 20 FM—plus 3 more per service if they are controlled by minorities and small businesses

LOCAL LIMITS

Old Rules	*New Rules*
Ownership of no more than one AM and one FM station licensed to the same market	In markets with 15 or more stations, up to 2 AM and 2 FM, as long as the combined audience market share does not exceed 25 percent. In markets with fewer than 15 stations, up to 3 stations may be owned, 2 of which may be in the same service area, as long as the number commonly owned stations does not exceed 50 percent of the number of stations in that market.

Note: The FCC defines a radio market as the area encompassed by the principal community contours of the overlapping stations proposing to have common ownership.

FIGURE 5.1

owned AM-AM or FM-FM stations in the same market to 25 percent of weekly broadcast hours. There remains, however, no restriction on same market AM/FM simulcasting.[8]

In announcing the new August 1992 radio ownership ruling, then FCC chairman Alfred Sikes said the commission had abandoned the "one size fits all" local ownership limits. In a prepared statement, Sikes said: "There is today a recognition that attaining and sustaining economic and thus programming health might well require larger holdings in bigger markets than in smaller ones. This conclusion, and the resulting deregulation, are at the heart of our action to provide radio station owners with more freedom and listeners with better programming."[9]

> This consolidation (of ownership) is good for the industry. It returns broadcasting back to the broadcasters and provides an exit route for a lot of banks and other operators. It also provides a buying pool that hadn't been there before.—Ed Christian[10]

As part of this increased deregulation in radio ownership, the FCC said it would examine investment incentives where radio group owners would be allowed to exceed national ownership limitations by a defined number of stations if they established and implemented a "broadcast ownership incubator program" designed to provide management or technical assistance, loan guarantees, and/or training to minorities and small businesses seeking to enter the radio field.

The FCC's 1992 radio ownership ruling catapulted the relatively obscure business term for multiple ownership, *duopoly*, into the limelight. The government's use of this term for multiple common-market station ownership will be long identified with the business and management side of radio broadcasting well beyond the twentieth century.

Actually, the use of the term as it applies to radio ownership is a misnomer. In the true sense, duopoly means two (and only two) entities having "preponderant influence or control" over an area such as an industry, community, or country.[11] New radio professionals, nevertheless, will undoubtedly hear this term at every stage of their radio broadcast careers.

At the same time the new duopoly rules were being implemented, station trading prices (excluding the largest radio markets like New York and Los Angeles) began to correct themselves with market values falling more in sync with station cash flow performances. The new ruling to expand station ownership allowed the industry to become more financially active.

Although there were not an overwhelming number of station sales applications in the initial weeks of the new rules, which some government officials feared might happen, after the first 18 months, the FCC reported nearly six hundred duopoly-oriented sales transactions.

One FCC official characterized this amount as a "steady influx of sales activity." Due to the high revenues typically generated by FM stations, many radio companies were doubling their FM ownership in markets from one to two stations. According to a mid-1994 analysis of the new duopoly radio marketplace conducted by Duncan's American Radio, over 28 percent of 12-and-over radio audiences were controlled by duopoly stations and just under 40 percent of radio revenues were going to duopolized stations. The radio industry was indeed taking advantage of this new duopoly ruling.

Radio station trading data provides evidence of heightened acquisition activity. According to the station sales figures compiled by Paul Kagan Associates, there was approximately $2.4 billion transacted in radio station sales in 1993 compared with $1.174 billion in trading activity for 1992. (See Figure 5.2.)

The expanded radio ownership model for markets with fifteen or more stations provides the backdrop for many large group deals, establishing new national mega-groups. In essence, the FCC duopoly ruling has driven up the competitive stakes in radio, causing most station acquisitions to be done purely from a market-specific stance. This means strategically building a strong station group in individual markets, particularly in those markets where a company already owns radio properties. In this way, a significant radio audience concentration could be achieved. Conversely, duopoly expansion creates the potential for increasing financial pressure in operating stand-alone or nongroup-owned AM and FM stations, especially those stations without a long-established audience following. Industry analysts, however, agree that the resulting shakeouts and underlying advantages

SUMMARY OF RADIO STATION TRADING FROM INITIAL DEREGULATION THROUGH DUOPOLY EXPANSION

Year	*Dollar Volume of Transactions*
1993	$2,370,000,000 *
1992	1,174,000,000 *
1991	742,000,000 *
1990	791,000,000 *
1989	2,147,000,000 *
1988	3,411,000,000 *
1987	2,652,000,000 *
1986	3,059,000,000
1985	2,258,000,000
1984	1,258,000,000
1983	1,272,000,000
1982	602,000,000

SOURCE: Paul Kagan Associates. Copyright© 1993, Paul Kagan Associates.

NOTES: The chart begins during the radio industry's initial ownership and programming deregulation phase—1983 being the first full year after the FCC's deregulation order—showing the incremental increases through the mid-1980s. It shows total proposed sales volume, including construction permits and partial sales. AM-FM combination station sales count as one transaction. *Canceled deals were omitted from the figures for these years.

FIGURE 5.2

from duopoly ownership will be in an evolutionary state throughout the nineties.

There are several positive economic indicators for radio as it progresses toward the next millennium. According to an analysis of market growth potential for the commercial U.S. radio industry prepared by Broadcast Investment Analysts (BIA), there will be 5 percent to 7 percent growth in the top 50 radio markets through at least 1997. (In 1991, these markets had a combined negative market growth rate of –1.7 percent. In 1992 the markets registered only 0.8 percent growth. (See Figure 5.3.)

The BIA analysis, which studies the relationship of local radio revenues to local retail sales, also shows a number of radio markets of varying sizes in the Sun Belt regions, including Jacksonville, Fla.; Austin, Texas; Phoenix, Ariz.; and Las Vegas, Nev., all poised to

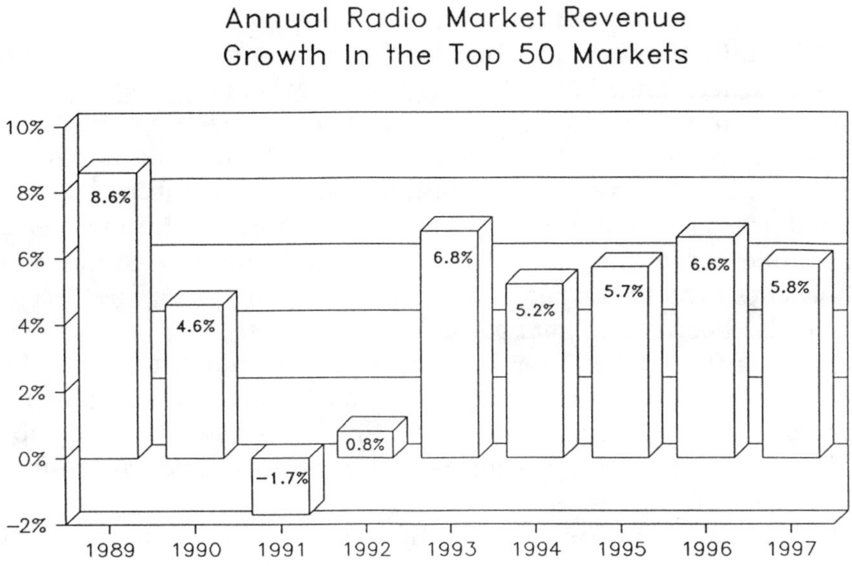

FIGURE 5.3

have annual market growth rates just over 7 percent through 1997.[12]

Another future byproduct of expanded duopoly situations is that the FCC ruling may revive interest from large multimedia conglomerates in radio ownership. Many major companies, including those that left the business during the mid-1980s because of the perception that there was not enough room in radio for them to be major players compared to other media industries, may decide to invest or reinvest in radio.

LOCAL MARKETING AGREEMENTS

Within this new radio landscape new business alliances are being forged among former competitors. As noted, when discussing AM radio in Chapter 4, local marketing agreements or LMAs began in earnest during the early 1990s. They have since become commonplace, providing better opportunities for local radio operators—AM and FM alike—to focus their time and money on competing with other forms of media instead of other radio stations. The LMA altered the economics of daily business.

Essentially, these time-brokered agreements allow two separately owned radio stations in the same market or region to jointly enter a programming and/or advertising agreement with the expectation that the LMA will mutually benefit both parties. Usually, a financially strong station reaches a combined operational and sales deal with a financially troubled station owner in the same community to oversee programming and commercial time sales for a fee or percentage of advertising sales. These deals have also been called *joint operating agreements*.

While the parties exercising an LMA are not required to file with the FCC, the licensee of the station that has entered a partnership to have its programming or air time leased is still responsible for meeting the station's control, maintenance, and community service requirements.

There is one major caveat to these kinds of programming and advertising arrangements among stations, however. If a station brokers more than 15 hours of programming per week on another station in the market, that station will be counted toward the brokering station's new permissible ownership totals.

In the 1990s, LMAs have served as a prelude to station ownership transfers. Essentially, two station owners in the same market could enter such a partnership arrangement with the major party holding an option to buy the station being brokered within a certain time frame.

Another aspect to LMAs occurs once a station sale agreement has been consummated between two owners in the same market. Here, the two parties may opt for an interim LMA deal giving the new owner a smooth transition for management, and especially programming. In this way, the LMA arrangement acts as a transfer buffer for the new owner's management assisting in the oversight of the acquired station. If knowledge of a pending format switch on the acquired station is discovered during the sale approval period, which can take a few months, the LMA could deter a competing station in the market from airing that format first.

This strategy, which is part offensive and part defensive, allows a potential programming franchise to be developed that complements the audience demographics of the new owner's other station or stations in the market.

There are, however, several potential antitrust pitfalls that radio owners should be aware of in any LMA arrangement. According to a legal paper on LMAs prepared for the National Association of Broadcasters and presented during the association's annual multimedia convention and exposition in 1993, collaboration on these areas should be avoided when structuring LMA deals:

- what rates can be charged on credit, payment, or any other terms of advertising sales;
- division of customers or otherwise allocating business;
- which geographic areas to compete in; and
- boycotting a third party.[13]

IMPACTING PERSONNEL

New local market duopolies, national ownership expansion, and sales and programming alliances come with different sets of operating rules that significantly affect radio station personnel, causing employment ranks to shrink. Due to the consolidation resulting from these deals, some station personnel—particularly upper management and programming and sales executives—are being displaced and/or relocating to smaller markets.

Although selection of the right duopoly employment formula will change throughout the 1990s, many consolidating radio companies are operating with separate managers as they acquire additional stations in a given market. Some station owners, however, choose to have one general manager oversee the operations of all commonly owned stations in a given market. In essence, they have become local market group operators.

Many local station groups are also only employing one sales manager and one program director for all stations. Often, the survivors are usually those executives from the acquiring company.

Joe Sullivan, president of Joe Sullivan & Associates, a New York–based management consulting firm specializing in executive search and recruitment for the broadcasting industry, described the early impact of duopolies on station personnel this way:

> The initial effect of duopolies has been to strengthen the position of some radio companies (through consolidation) in various markets—

> more at the larger and major markets than at smaller markets, but a trickle-down effect will occur. As duopolies occur, people at the higher end of station management are being released . . . . But good sales talent is always hard to find. A good general sales manager who loses out in a large market can find a general sales manager's job in a medium-sized market.[14]

As for maintaining separate sales forces for same market duopoly stations, here too, depending on market size, one could find an equal number of owners who say they have a separate staff as those who have consolidated.

But what some major market radio owners find is that the decision to operate separate sales forces for new multiple station structures—especially for two FM stations in a single market—has been cost-effective for selling the *strengths* of the individual stations in the local group.

Infinity Broadcasting followed this path. Mel Karmazin, in a *Radio Ink* magazine profile, explained the company's views on station personnel consolidation as it relates to duopoly ownership situations:

> Radio has a high degree of fixed costs and relatively low variable costs. Once you get beyond your fixed costs, roughly 80% in excess of that falls to the bottom line. Therefore, if you're able to generate a million dollars in incremental revenue, you can increase your cash flow by $800,000. We believe that by having separate staffs and general managers, we can generate incremental revenues much greater than the savings we could realize by cutting a salary.[15]

During one of the first organized industry discussions of the potential impact of the FCC's expanded ownership ruling on radio careers, Diane Sutter, executive vice president/operations (radio and television) for Shamrock Broadcasting, owned by the Disney family, underscored the need for radio station programmers in the duopoly age to thoroughly understand their local station group's *sales and marketing strategy*. (The meeting took place at the National Association of Broadcasters' 1993 multimedia conference and exposition in Las Vegas.) According to Sutter:

> Versatility is paramount. The program director has to understand that the rules have changed and that when rules change nobody has

an advantage and that puts everyone back to zero . . . . We have a situation in San Francisco where we have four stations each with separate programming. That is a very new situation for everyone involved and managing that situation is going to take real skill and talent. Not only understanding the programming, but understanding how to manage the dynamics of that situation is essential . . . .

The ability to lead a staff through the transition process and into the new duopoly environment will be a test of leadership skills. The program directors who demonstrate those leadership skills will be the ones there the following year.[16]

DEMOGRAPHIC STRATEGIES

Some marketers are still discovering something that many radio broadcasters have known for some time: all consumers, just like all listeners, are not alike. In this information-laden age, products need to be tailored to individual tastes. Matching product to consumer needs is imperative for any industry to survive through the 1990s and beyond.

In the automotive world, for example, the industry studied changing consumer tastes and saw that the baby boomers, who have traditionally dictated tastes in this business, wanted cars customized to their new needs, including safety features, good engine velocity, reliability, and a contemporary look at affordable prices. Therefore, General Motors introduced the new Saturn product line and a redesigned Geo Prizm product line to satisfy these demands.

Radio broadcasting has also customized its programming to meet the evolving music and information tastes of its listeners. It is the reason the medium has become so highly specialized. The one inherent aspect to radio is that it has always been a reflection of society and culture.

With the FCC's expanded duopoly rules comes the opportunity for local radio stations to refocus their positions both *vertically* and *horizontally*. Some broadcasters, for example, will try to vertically dominate the 18- to 34-year-old demographic market by having one FM format geared to 18- to 34-year-old men and the other FM format geared to 18- to 34-year-old women—such as modern or progressive rock and an adult contemporary format—or both geared to one gender within a specific demographic.

On the other hand, a broadcaster could penetrate the market horizontally by programming one FM station to attract the 18- to 34-year-old market and the other FM station for the 35- to 54-year-old market. There could be two versions of the same format, such as a young-appealing country format with more traditional country sounds. And on a commonly owned AM station in the same market, the owner could program formats for ages 54 and above, such as a nostalgia or big band format.

If a radio company has two of the top 5 stations within a certain format or demographic bracket and/or covers the 25- to 54-year-old spectrum, it has a very dynamic audience for both local and national advertisers. With more listeners now attracted to FM stations, owning two FM stations in the same market makes this an easier feat to accomplish than under the previous FCC ownership rules. Under certain market conditions, these stronger stations could also drive up higher average unit rates—prices for commercial spots—which, in turn, directly impacts cash flow and profitability.

The key here is to establish format synergies. Unless a radio owner controls a certain demographic in the marketplace with multiple stations, advertisers and buyers of media time may still opt to buy time at the one station in the local group they perceive will give them the desired demographic and gender audience for their product or service.

Bud Paxson, chairman and CEO of Paxson Communications, a Clearwater, Fla.–based broadcasting company, has a number of duopoly situations in different markets. He compares marketing multiple radio stations in the same market with complementary formats to marketing a local newspaper with varying sections—that is, representing radio to advertisers as "one audience made up of various parts."

According to Paxson, a one-time radio broadcaster and former president of cable's Home Shopping Network, the way a local newspaper views circulation as encompassing all of its different sections, such as business, sports, and classified advertising, is the way radio owners should think of the weekly audience cume of their local station group. This would require a team specialist selling approach.[17]

Following this analogy, radio advertisers could then choose to reach the entire audience spectrum of the duopolized stations or any particular segment of that audience.

NATIONAL PROGRAMMING SHIFTS: THE SYNDINETS

Besides new demographic strategies, the industry in the 1990s saw a new shift in programming strategies—to *morning radio syndication,* which also became part of radio's realigned structure. Many radio stations across the country are airing nationally distributed morning music- and talk-radio shows. Star performers or personalities are featured in an entertaining format during the morning drive-time period—traditionally the time that attracts the most listeners with local news, sports, weather, and traffic reports. As discussed in Chapter 1, for this reason it has been the one daypart within a radio station's programming day that station management scrutinizes.

What caused this change to the programming structure of the medium is a more concentrated results-oriented approach to ownership, mostly due to radio's high debt service years. Owners and managers only had to witness improved ratings performances on the initial stations that subscribed to syndicated morning radio shows to see that this had fast become a proven direction for them to follow. With access to a pool of increasingly visible national morning radio talent that expanded rapidly in the mid-1990s, station management could attract more listeners.

At the same time, they could keep costs down in the new duopoly era by not having to employ their own full-time morning personality—typically the highest paid air shift. According to Walter Sabo, president of Sabo Media, a New York–based radio management and program consultancy:

> This kind of network programming is good for radio. It raises the high watermark for radio by raising public awareness of the medium . . . . And it frees up some significant dollars to invest in talent during the rest of the day.[18]

This rise in locally produced, syndicated morning programming, part of the trend known as the *SyndiNets,* should come as no surprise to students of radio. After all, programming is radio's basic product. Even the early network radio programming executives all follow one edict: "If it's quality programming, they [listeners] will come." One need only read the mission statement of the industry's first commercial radio network, NBC

Radio, when it was formed by RCA in 1926. The purpose of NBC, RCA said, "will be to provide the best programs available for broadcasting in the United States" and to offer them to stations "so far as it may be practical to do so and they may desire to take them."[19]

When blocks of network radio shows transferred to television in the 1950s, network radio programming was reduced to regularly scheduled news and special event programming, sports, and a few variety shows, such as Don McNeill's *Breakfast Club* based in Chicago, which ran on the ABC Radio Network each weekday morning through 1968.

The next generation of radio network entertainment shows surfaced in 1970 when Los Angeles-based announcer Casey Kasem and Watermark, a Los Angeles radio program production company, teamed up to launch a weekly Top 40 countdown show, called "American Top 40," relying on *Billboard Magazine*'s pop charts. The show was distributed by ABC, which at the time owned Watermark. (Kasem left ABC in the late 1980s to host countdown shows and other programs for Westwood One).

As the proliferation in radio programming intensified in the seventies and early eighties, the Kasem countdown show spurred a series of countdown radio programs and other specials mostly geared for young adult audiences. The hosts were highly talented local personalities, many from Los Angeles, who were transformed into national radio personalities. Other shows were hosted by already-noted radio entertainers such as Gary Owens, who had appeared on NBC-TV's *Laugh-In* show in the late 1960s, and Murray "the K" Kaufman, a popular radio personality in the early days of rock 'n roll.

Most of these programs were taped during the week for weekend airplay. Before the satellite-distribution era, the shows were usually put on discs and mailed to the station, complete with national commercial spots, just before the scheduled weekend of airing.

These new breeds of nationally produced weekly radio shows, syndicated either through major networks like ABC or NBC or independent production/syndication firms, were in effect ad hoc or special program networks unto themselves.

Special weekend shows are still offered by major radio networks and program suppliers, but the new national syndication trend in morning radio has moved nationally distributed programming into a new realm for modern-day radio owners. Locally produced morning radio personalities, notably Infinity's Don Imus and Howard Stern, heard over all-sports WFAN-AM New York and classic rock WXRK-FM New York, respectively, are distributed to local stations around the country. Both are high-profile, nationally known, controversial talk personalities.

It was Infinity's decision in 1986 to air Stern's WXRK program on its Philadelphia station, WYSP-FM, that started the trend. While industry observers doubted that a local New York radio show would work in another market, Infinity and Stern proved the skeptics wrong. The philosophy behind this move was simple: listeners will be attracted to good radio talent—generally comedic in nature—in the same way TV viewers are attracted to Johnny Carson or David Letterman.

During its first two years in Philadelphia, Stern's morning show gained considerable ratings share and became a huge success. Infinity's move marked a return to national personality radio.

Infinity then aired the show on its radio station in Washington, D.C. WJFK-FM, and in 1991 took Stern's show beyond the Northeast to KSLX-FM Los Angeles, a non-Infinity station owned by Greater Media. There, too, significant ratings were recorded. In the summer of 1992 Stern emerged as the highest-rated morning radio personality in the market.[20] Infinity showed the radio industry that syndicated morning programming—with the proper talent for the format—does work. (See Figure 5.4.) Stern's broadcast has since been cleared on a number of stations nationwide, including most of the top 10 radio markets.

Shortly after Infinity assumed control of WFAN-AM New York in 1992, the company syndicated the highly popular *Imus in the Morning* program, as well, distributing the broadcast through its newly operated Unistar Network division. Other morning personalities soon followed suit. By 1994 the number of new morning syndicated shows was soaring.

ABC Radio, for example, launched both *Moby in the Morning*, which originates from the CapCities/ABC–owned WKHX-FM Atlanta, and *The Tom Joyner Show*, originating from the ABC/Satellite Music Network studios in Dallas. The latter is designed to be a nationally originated broadcast from a major studio with its own in-house band. The Premier Radio Networks, a Sherman Oaks, Calif.-based radio program supplier, began distributing

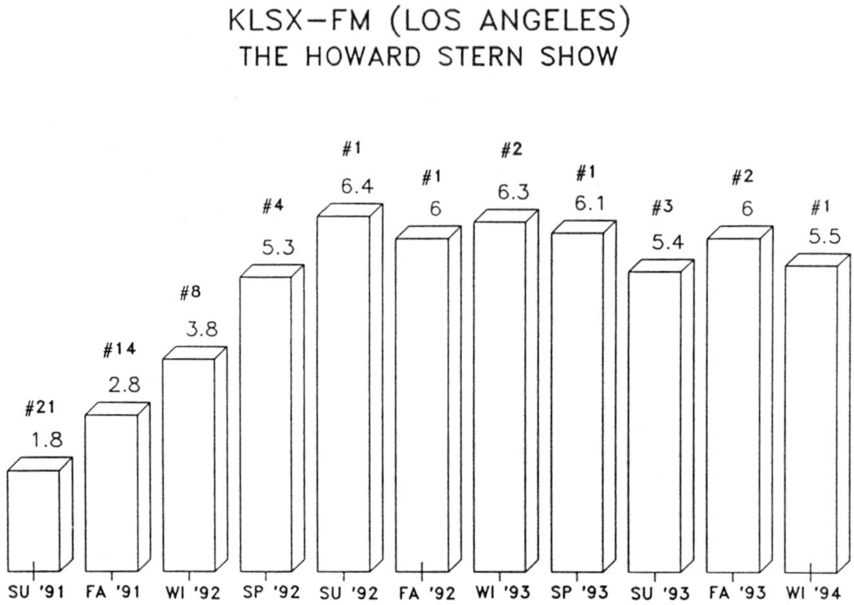

FIGURE 5.4. The illustration above charts the strong ratings performance of *The Howard Stern Show* on KLSX-FM Los Angeles from its debut there in the summer of 1991 through the winter of 1994. The figures represent Arbitron metro ratings for Monday through Friday, 6:00 A.M. to 10:00 A.M. for persons 12 years and older, in average quarter hour shares.

Gerry House & The House Foundation, which originates from WSIX-FM Nashville.

During this period of the 1990s, several locally produced, regional *morning team* shows were also distributed, mainly to stations in the South. Compatible formats for syndicated morning fare are usually music-oriented and, depending on the talent's origination, can vary from rock to country. In the view of Pat Welsh, vice president of music programming for the Pollack Media Group, a Pacific Palisades, Calif.-based media consulting firm:

> If you could bring among the more talented radio people in the world to your radio station, whether that person is broadcasting from your studio or coming from New York really doesn't matter. If people like it—they will listen . . . . And this also gives small and medium markets access to major talent.[21]

Besides distributing these programs, morning syndicators have calculated a way to alleviate station management's concerns about relinquishing local control of their morning drive-time programming. Within many of these broadcasts is a built-in hourly average "cut-away" time of 22 minutes in which local stations have the option to air weather, traffic, news, and other information. This allows the shows to be presented in a modular format, which national programmers reasoned, permits station owners to air quality morning talent and simultaneously localize the product. In this way, the morning show can provide both entertainment and information.

An integral part of this new SyndiNet radioscape are conservative talk-show hosts such as G. Gordon Liddy, and Rush Limbaugh, who has experienced the most success on AM talk radio in the United States since Larry King entered the arena in the late 1970s. (For more on Rush Limbaugh's success, see Chapter 4.)

Essentially, the SyndiNets are live programs airing within a defined daypart in much the same way programs were carried by network affiliates in the thirties and forties. Radio broadcast consultant Walter Sabo, who is also a former NBC and ABC radio programming executive, observes that the shift back to national personality radio, a major force in the remaking of radio, is a boon for attracting more young creative talent to the medium.

Some national programmers, meanwhile, say that the radio industry in general had not done a sufficient job of nurturing quality local talent, particularly morning drive-time personalities, for the country's 260 rated radio markets.

NOTES

1. "Outlook 1994," *Radio & Records*, Dec. 31, 1993, p. 24. Steve Goldenstern is executive vice president and group program director for Saga Communications
2. "59% of Radio Stations Lost Money in '91," *Broadcasting Magazine*, July 6, 1993, p. 6.
3. Based on an interview with Frank Osborn of Osborn Communications, New York.
4. Based on statement of FCC Commissioner Erwin S. Duggan, *Regarding Revision of Radio Rules and Policies* (Docket MM 91-41), August 1991.

5. Based on an interview with Jeff Smulyan of Emmis Broadcasting, Indianapolis, Ind.
6. "Sikes Sees Possible Revision of New Ownership Rules," *Radio & Records*, April 3, 1992, p. 4; "New Radio Rules Not Set in Stone," *Billboard Magazine*, April 25, 1992, pp. 6, 67, 70–71.
7. Based on FCC's *Report & Order on Radio Ownership Rules* (Modified on Consideration—Docket MM 91-141), August 1992.
8. Ibid.
9. Excerpt of statement by then FCC Chairman Alfred Sikes upon announcing modified radio ownership rules (Docket MM 91-141), August 1992. (National minority ownership limits for AM and FM were raised to twenty-five each in 1994.)
10. Based on an interview with Ed Christian of Saga Communications, Grosse Pointe Farms, Mich.
11. Definition of *duopoly* from *Webster's Ninth New Collegiate Dictionary*.
12. Data from Broadcast Investment Analysts' *Investing in Radio 1994*, BIA Publications.
13. "LMAs and the Antitrust Laws," paper presented by attorney Bryan Cave during the National Association of Broadcaster's 1993 Spring convention, Las Vegas, Nev.
14. Based on a conversation with Joe Sullivan of Joe Sullivan & Associates, Southold, N.Y. Joe Sullivan is also a small market duopoly owner of WBAZ-FM in Southold, NY, and WLIE–FM in Bridgehampton, NY.
15. Based on an interview profile of Infinity Broadcasting president/CEO Mel Karmazin in *Radio Ink*, Jan. 3, 1994, p. 21. Karmazin was named "Radio Executive of the Year" (1993) by the publication.
16. Based on panel discussion entitled, "Career Survival in the Age of Duopolies," National Association of Broadcaster's 1993 Spring Convention, Las Vegas, Nev. Sutter has since become president of Shamrock Television.
17. Based on an interview with Lowell "Bud" Paxson of Paxson Communications, Clearwater, Fla.
18. Based on an interview with Walter Sabo of Sabo Media, New York.
19. "The First 60 Years of NBC," *Broadcasting Magazine*, June 9, 1986, p. 50.
20. Data based on Arbitron ratings morning drive-time information for Los Angeles metro area; Monday through Friday; 6:00 A.M. to 10:00 A.M.; persons 12+; average quarter hour (AQH) shares.
21. Based on an interview with Pat Welsh of the Pollack Media Group, Pacific Palisades, Calif.

6

The Electronic Marketplace

Author's note: The following two chapters present an overview of developing applications in information and digital technology for radio as the industry approaches a new century.

MANAGING RADIO IN THE INFORMATION AGE

Automating the Processes

> Modern information technology increasingly combines functions to eliminate many conversion steps.—Joanne Yates and Robert I. Benjamin[1]

Today's fiercely competitive multimedia environment promises to broaden substantially during the late 1990s and well into the twenty-first century. Radio's viability as an advertising medium will be challenged if it does not adapt to the way sales information about advertisers is now processed—that is, the computerization of buying, selling, and tracking media. Many traditional buying and selling methods of advertising time are being automated.

The principal changes involve a move from person-to-person placement and scheduling of media time—conducting order placement and scheduling in person or by telephone—to a personal computer (PC) process with interactive components. It also involves high-quality, timesaving laptop computer presentations by the sellers of media time.

For radio, many of these new initiatives are taking place first on the national level—mainly in the national spot business arena. Essentially, what is being developed for radio advertising are compatible software systems for both sellers and buyers. One of the first compatible radio marketing software designs has been coined the AM–FM Buyer's Worksheet. The program was created through a cooperative effort of Donovan Data Systems (DDS), a New York City–based media services company, several leading advertising agencies, and the Interep Radio Store. The software program shows all of the quantitative criteria on which radio is evaluated, including target demographics, station ranking by market, pricing, and scheduling. It also includes qualitative factoring capabilities to avoid buying purely by numbers. The concept here is that the buyer and seller of radio time work from the same software design. This consistency allows radio sales presentations to be prepared in the same computer format used by agency executives.[2]

At many large media companies, where advertising and research information was once stored on a mainframe computer, there are now workstation PCs that instantly access both quantitative and qualitative data. Also important in the area of information technology is the move to electronic invoicing systems by most major advertising agencies.

Any media organization with a PC information technology (IT) strategy, whether it is on the agency or broadcast/cable side of the business, will also understand that this technology does not replace manpower, or what corporate executives call *salespower*. For marketing radio stations, personal interaction is still necessary between radio salespeople and advertising executives when discussing the characteristics of a product or service and the stations being considered for the buy.

Information technology, however, allows station sales executives to more efficiently meet the needs of the advertisers in matching audiences to products, creating a higher productivity output. With an automated advertising placement structure, sales assistants can focus their efforts in ways that are more beneficial to their company or station.

A 1993 report on new technologies in *BusinessWeek* states that "history offers striking evidence that technological revolutions eventually create many more jobs than they destroy," further showing that new job skills result from technology advancements.[3]

The implementation of a PC information technology strategy, which usually involves the creation of local area networks, eliminates time-consuming paperwork with its high potential for human error. These networks allow sales personnel to immediately respond to research requests and to access other records and files. It also automates the scheduling of various commercial campaigns. For radio, managing sales with information technology will only advance the medium's marketability. It has emerged as a mandatory "tool of the trade." As we have seen in previous chapters, technology is the cornerstone of progress. In their paper on information technology for modern-day businesses, Joanne Yates and Robert I. Benjamin write:

> By lowering costs and speeding processes, IT improves firms' profitability and competitiveness. More significantly, though perhaps less commonly, especially in the historical period, IT has also been used to support and even drive new strategies.[4]

THE LOCAL RADIOSCAPE

For any service-oriented business that unites suppliers and buyers of a product or service, a transfer from manual routines to automated information technology systems must occur in order for that business to effectively compete with other industries. As mentioned above, this transfer drives new strategies. Since the late 1970s, local radio stations have gradually automated their routine traffic and billing systems. With the 1990s, however, comes the installation of PC networks and dedicated marketing software, which help radio station owners establish a more effective pricing structure. This application of information technology to inventory control management is known in corporate corridors as *yield management*.

Yield management first began in the airline industry during the 1980s. This strategy strives to set the current optimum price for inventory, based upon historical data. For airlines, that would be available seating; for a radio station (and also a network) that would be available commercial units or spots. Yield management software gives executives more data with which to make an intelligent decision to lower or raise prices as market conditions change. It helps correlate available inventory with pricing to achieve the highest possible sales revenue.

In the electronic information era, yield management can also be an instant barometer on current sales and predictions for upcoming months. One such system, called *Market Sellout Percentage,* has been developed for local radio stations by the accounting firm Miller, Kaplan, Arase & Co. It calculates the number of commercial station inventory units available as a percentage of the total in a specific market against the number of commercial inventory units sold. (See Figure 6.1.)

Among the different variables determining the maximum prices for commercial broadcast inventories are the probability advertising customers will pay the quoted price based on the current demand; the possibility the same spots could be sold to another advertiser for more money; the length of time until the spots have to run—that is, the customer's urgency to get on the air; and the amount of unsold spots in relationship to the amount of time left to sell them.[5]

Within a demanding advertising arena, it has also become imperative for radio station executives to establish account management computer models to track and monitor the value of their sales contacts—day-to-day transactions, including prospecting, analyzing, and closing accounts. This gives the owner and/or manager the opportunity for instant analysis and intervention.

One method of account management comes from New York City–based Competitive Media Reporting (CMR). Their quarterly local market radio spending analyses show subscriber stations their total dollars for each account in the market and also their share and rank for these accounts. (A similar report is done for national representation firms, showing their account standing in key market centers.) These advertiser tracking reports, which depend upon confidential submissions of advertising revenue by individual stations, began in several major cities in the early

Market Sellout Percentage *Flash Fax*

To: Station General Manager
Fax Number: (800) 555-1212

Market Sales Pacing Report
As of February 15, 1994

	Number of Units Available for Sale	*Number of Units Booked*	*Sellout %*	*Station Rank*
FEBRUARY				
Market	60,480	48,142	79.6%	—
Station	5,040	3,933	78.0%	3
MARCH				
Market	75,600	40,219	53.2%	—
Station	6,300	2,966	47.1%	2
APRIL				
Market	60,480	17,358	28.7%	—
Station	5,040	883	17.5%	4

The above sellout factor data was compiled based upon reports from the following 12 stations: KAAA, KBBB, KCCC, KDDD, KEEE, KFFF, KGGG, KHHH, KIII, KJJJ, KLLL, KMMM

MILLER, KAPLAN, ARASE, & CO. CERTIFIED PUBLIC ACCOUNTANTS

FIGURE 6.1

1990s. This kind of management data can help station executives determine the amount of business being generated on a seasonal basis. CMR also offers local multimedia reports showing account spending across several measured media formats.

Local stations can now electronically receive national spot contracts and information about specific buys sent from a national rep's computer via telephone lines to their fax machine or computer with a fax modem. Eventually, this information will go directly into a station's computerized traffic system.

Meanwhile, local major radio networks affiliates can now receive network commercial feeds via satellite. Scheduling instructions to network commercials are also being sent by satellite. At ABC Radio, for example, in the late 1980s the network division formed a data transmission unit, called ABC Data, that sends commercials via satellite with accompanying scheduling instructions and special program information to its affiliates. The data signal is then decoded at the station and either appears on a printer or is directly sent to a computer.[6]

By the new century, there should be fully addressable, interactive information systems—including electronic mail (E-mail)—in place between local radio and their national representatatives and network partners.

It is important to note that the role of information technology for the present-day radio industry should not be overstated or understated. New mechanisms are necessary to properly channel the ever-increasing amount of data available. The speed by which this information is processed will lead to greater efficiencies in the marketing of radio.

NOTES

1. From "The Past & Present as a Window on the Future," taken from *The Corporation of the 1990s*, (Oxford: Oxford University Press, 1991), a compilation of writings on information technology. Joanne Yates is senior lecturer and coordinator of management communications at the Massachusetts Institute of Technology's Sloan School of Management. Robert I. Benjamin is a visiting scientist at the Sloan School.
2. Based on information from Bob Lion, regional executive/technology director, the Interep Radio Store, New York.
3 "The Technology Payoff," *BusinessWeek*, June 14, 1993, p. 72.
4. "The Past & Present as a Window on the Future," *The Corporation of the '90's*, Michael S. Scott Morton, ed., (Oxford University Press, 1991), 81.

5 Leonard M. Lodish, "Applied Dynamic Pricing and Production Models with Specific Application Broadcast Spot Pricing," *Journal of Marketing Research*, the American Marketing Association, May 1980, pp. 205-206. Leonard Lodish is a professor of marketing at the Wharton School of the University of Pennsylvania.

6. Based on information from Jeff Mathieu, manager of Affiliate Technical Information, ABC Radio Networks.

7

The Digital Factor and Other Technological Applications

If all human communication is becoming digital and the consumer has more microprocessor-based devices and we digitize the transmission path, who says that radio must be limited to sound?—John Abel[1]

RADIO BROADCAST DATA SYSTEMS: EASING THE SELECTION PROCESS

On the heels of radio's emerging entrepreneurs comes emerging digital broadcast technology that promises to revolutionize the business.

In the mid-1990s, the U.S. radio broadcasting system is adapting its inherent structure of subcarrier frequencies (unused portions of main channel assignments) for digital data transmissions from station to receiver. The use of subcarrier frequencies by radio is not new. Historically an FM application, stations on the FM band have been leasing their subcarrier channels for supplemental services, such as background music programming, data, and paging services to outside suppliers since 1955. These leasing arrangements have provided FM broadcasters with additional revenue.

The transmission of digital text—the zeroes and ones of binary computer language—direct from local stations to auto and home receivers is a more recent consumer development, however. This technology is also known simply as *radio broadcast data systems* or by the acronym *RBDS*.

The technology allows station operators to transmit data information to listeners over specially equipped "smart" radio receivers, which can perform a variety of automatic tasks. This information can display a station's call letters, logo, and format and also identify songs and artists. RBDS technology gives listeners more flexibility in selecting stations by making radio tuning easier.

RBDS receiver manufacturers will concentrate initially on car radios. The first automotive RBDS receivers are being designed with data display panels that receive up to eight alphanumeric characters. These receivers will also have scanning functions to allow a listener to select a station by format. (Through timed computer signals, broadcasters whose formats are made up of combined programming elements can change a station's format identity by daypart.)

According to a National Association of Broadcasters paper on RBDS technology, radio executives using this technology could program a radio to automatically switch to an affiliated station carrying identical programming if a listener travels out of the main signal's range. In this way, "a station could 'hand-off' its signal to another station with which it has a simulcast (or LMA) agreement."[2]

Another important RBDS function is its ability to identify a station as one that provides traffic information so that when listeners want such information, an RBDS automotive receiver would quickly scan the dial for that station. This real-time traffic information could also switch on automatically when the receiver is in a waiting reception mode and its audio signal is muted. (See Figure 7.1 for an example of an automotive RBDS receiver).

Aside from its advantages for car receivers, there are more possibilities for RBDS home receivers, which could either come

FIGURE 7.1. Denon DCR-730R

equipped with a large display panel or be attached to a monitor with scrolling text capabilities. In this way, a variety of data could be transmitted.

RBDS transmissions will buttress the positioning of local radio stations in their individual markets while providing access to previously untapped sales and promotion marketing revenues.

The RBDS technology also lends itself to what can best be described as "direct-result" advertising. An example is a proposal called *CouponRadio.* This adaptation of RBDS technology is designed initially to deliver special electronic coupon information to listeners through their car radio receivers.

Along with the audio portion of a commercial, text information (or infomessages) regarding the advertised product or service can be transmitted by the station. If the listener is interested in the advertised product or service, the listener can press a button on the radio to automatically store the information transmitted on a *smartcard* or *radiocard*—a device the size of a credit card that contains a memory chip. Also on this card is the time, date, and call letters of the station the listener was tuned to when storing this information.

These radiocards are then taken by the listener to the participating advertiser and/or retailer. Here they are processed through a CouponRadio terminal box or kiosk programmed to instantly search the card for an advertiser's specific coupon infomessage, including discount offers. With CouponRadio, founded by New York–based entrepreneur David Alwadish, the artist and title of a song can also be stored. For radio station management, this interactive use of RBDS technology allows for closer monitoring and measurability of advertising and promotional messages.

Effectively, CouponRadio's RBDS technology merges real-time broadcasts with computer storage. Besides acceptance by the radio broadcasting and advertising communities, what is necessary for CouponRadio to become a reality is the development of RBDS radio receivers with computer memories.

Digital data transmission technology for radio listeners has been used in Europe since the 1980s—known there as radio data systems (RDS). It wasn't until January 1993 that the National Radio Systems Committee (NRSC) issued an official U.S. standard for RBDS transmission: the FM subcarrier frequency of 57 kHz. (A description of the NRSC appears in Chapter 4.)

This is the only standardized digital subcarrier system in the United States. Tests are being conducted by the NRSC to see if a similar RBDS standard (or other provisions) can be adopted for AM stations.

To implement RBDS technology, FM station owners and managers must purchase special digital encoders and input specific RBDS codes, which are then used to activate specific RBDS receiver functions. As of the mid-1990s, costs for these encoders were averaging $2,500 to $3,000.[3]

Among the companies developing RBDS radio receivers are Denon America (see Figure 7.1), Delco Electronics Corp., and Blaupunkt. Costs for the first RBDS automotive receivers carried an average premium of $50 over traditional radio receivers.

DIGITAL AUDIO BROADCASTING

A method by which radio stations could deliver compact disc–quality, digital audio broadcasting (DAB) directly to receivers is being thoroughly examined by both the radio and electronic industries.

Most systems being studied use an in-band, on-channel model, known by the acronym *IBOC*, for DAB broadcasting. IBOC allows the station to transmit a digital programming signal underneath its main terrestrial analog signal to a DAB radio receiver that can extract the digital signal. In theory, stations with a DAB in-band, on-channel license could duplicate their existing programming off the main channel or even program a separate format. There are still many unanswered questions, however, such as potential DAB interference to the main program channel or existing service area.

The IBOC systems being examined are primarily for FM channels. At least one company, USA Digital, a consortium comprised of CBS, Gannett, and Westinghouse engineers, is studying this technology for both AM and FM stations. Their examination has been dubbed Project Acorn.

DAB satellite signals, proposed by several companies, have the potential for mobile and portable reception, for example sending digital programming directly to a car receiver. This would spawn a series of new national programming services competing with local radio operators.

In this scenario, DAB programming would have to operate within an entirely new frequency spectrum. This has caused many radio station owners to be skeptical of any widespread acceptance of such a broadcast structure, noting that it could take years to build listener and advertiser support. While radio industry executives have adopted a wait-and-see attitude about DAB in general, they also believe that existing licensed broadcasters should be involved with any new DAB development, both to ensure proper phase-in and to reap any potential benefits.

With DAB comes new configurations of radio receivers for program reception along with mass acceptance by both the radio broadcasting industry and the public. If radio history is any indication, new technological developments for program delivery take at least a decade or more to move from the experimental stage to implementation. During the early days of FM broadcasting, for example, *The New York Times* in a 1939 editorial said, in part, that changing radio sets to accommodate new FM stations "was almost like changing the gauge of railways to accommodate wider and more luxurious cars."[4] AM stereo, originally developed in the 1960s, is another example. The media industry has also seen new forms of technology restructure entire segments, such as the introduction of compact discs and the demise of the vinyl record industry.

DIRECT (TO HOME) BROADCASTING AND CABLE AUDIO

While the technology is already in place in other parts of the world, such as Europe and Japan, the ability to send digital video and audio signals directly from a programming source to the home or other stationary location, via satellite, has become a reality in the United States. This technology is known as direct (to home) broadcasting or more aptly, direct broadcast satellite (DBS).

On December 17, 1993, a three-ton satellite for direct broadcasting, dubbed DBS-1, was launched by GM Hughes Electronics on an Arianespace rocket from French Guiana. There are two companies initially planning to provide high-powered DBS services: GM Hughes' own DBS service, called DirecTV, and Hubbard Broadcasting's United States Satellite Broadcasting, the latter leasing transponder space on DBS-1 from GM Hughes. Because the main thrust behind DBS is to provide video programming, this technology is better poised to serve the broadcast and cable television industries as they enter the twenty-first century.

Some commercial broadcasting companies, however, are exploring the potential for digital audio delivery via DBS. One early entrant is WBBR-AM New York. The station, whose call letters stand for Bloomberg News Radio, programs a business-oriented format featuring news and financial information. The format, implemented by Wall Street financial executive Michael Bloomberg, is available nationally as an audio service on DirecTV.

WBBR is part of a multimedia programming service offered by Bloomberg that provides both on-line computer business and financial data, a TV news service, and an audio package of the station's news reports.[5]

Regarding digital cable audio (sometimes called cable radio), these commercial-free subscription services have existed since the seventies, but only in the nineties have they become a business force. Despite the fact that cable radio—audio entertainment channels delivered by cable directly to a stationary location—is a pay service, national in scope, its effect on traditional U.S. radio listenership has been minor. Most cable audio services charge subscribers from $7 to $10 per month.

Recently, digital cable audio went beyond traditional cable distribution to increase circulation by taking advantage of direct-to-home broadcasting. Digital Cable Radio, a Horsham, Penn.-based digital audio service, began supplying its program to DirecTV. To position itself as more than cable, the company changed its name to Music Choice.[6]

These digital audio signals are delivered, along with digital video, directly over the DBS satellite to subscribers with an RCA-developed satellite dish. These consumer dishes, priced at about $700, are being built by Thompson Consumer Electronics. The unit includes a small 18-inch dish antenna (compared to the 10-foot backyard dishes seen in many rural areas), an integrated receiver/decoder, and remote control.[7]

DIGITAL PHONE TRANSMISSIONS

For both radio network and local station remote (on location) radio broadcasts, digital, point-to-point program transmission systems via dedicated telephone lines became widely accepted in the early 1990s. In essence, this technology allowed the telephone industry to reenter the radio audio transmission business.

This system, known as the Integrated Services Digital Network (or ISDN), is a high-fidelity link (sometimes fiber-based) that is slowly supplanting satellite technology for transmitting on location special events programming, such as coverage of the Olympic Games or a live concert, back to a network. In the 1980s, a portable satellite uplink dish was used to send this kind of programming back to a central network facility for nationwide distribution. Major radio networks and national program suppliers, however, continue to choose satellite technology for simultaneous feeds direct to affiliated stations nationwide.

ISDN technology is also being used by an increasing number of large major market stations for remote broadcasting from both distant and nearby sites, including local sporting events.

Changes in digital broadcast technology are occurring enormously fast. Radio station owners are automating the main broadcast and production studios with digital technology. This involves installing PC-based audio equipment for on-air compact disc play as well as for storage of commercial spots.

As with the information technology systems discussed in Chapter 6 and their relationship to managing the advertising processes, the computerization of studio functions will provide both station talent and program directors a more efficient workplace for managing the programming processes.

In essence, digital technology will eventually facilitate the integration of all radio station functions.

NOTES

1. Based on information from John Abel, executive vice president, National Association of Broadcasters.
2. Based on a paper entitled "An Overview of the Radio Broadcast Data System," prepared by John Marino, manager of Technical Regulatory Affairs, National Association of Broadcasters' Science and Technology Department, March 1, 1993.
3. "Broadcasters Tune in to RBDS for New Revenue Potential," *Billboard Magazine*, June 5, 1993, pp. 1+. Also based on information from John Marino of NAB.
4. Tom Lewis, *The Empire of the Air: The Men Who Made Radio*, (Edward Burlingame Books, 1991), 273–74.
5. "WBBR Is First Audio Superstation," *Radio World*, March 23, 1994, pp. 1+.
6. "DCR to Begin Satellite DAR Service," *Radio World*, March 9, 1994, pp. 1+.
7. "Betting on Small Dish TV," *The New York Times*, Dec. 15, 1993, pp. D-1+.

8

Epilogue

THE LESSONS

For the first time since the early 1950s, when television took center stage, the 1980s saw the radio business was offered a second chance with destiny. This time, however, its leadership came away with even greater insights for building the industry.

Perhaps the major lesson the emerging entrepreneurial owners and radio executives took from this freewheeling time is a simple one: rapidly assembling a station group or network more on *debt* than on equity financing will eventually lead to a cash flow dilema, particularly when stations are faced with new competition. The greater the need to produce profits to service large debt payments, the greater advertising rates will be compromised to capture business. But short-term gains do not necessarily translate into long-term profits.

Unless there are inherent guarantees of market-specific revenue growth, only financially strong owners, or as we stated previously, those who master radio marketing and financing skills, along with the major market players who operate the country's more profitable stations, will ultimately emerge as victors.

Radio broadcasting is a good case study of the entrepreneurial corporate culture in America during the 1980s. It is a study of how a highly desirable business for investors—one with fixed operating costs and historically high operating margins—could fall on such hard financial times as it did in the early 1990s.

> Where many radio broadcasters fail is thinking of the medium as some kind of special art . . . . that does not live and breathe on finance.—Frank Osborn[1]

This leads to another vital lesson taken from this turn of events: History no longer applies. Radio station owners should continually approach radio management with an innovative set of business and marketing ideas. "Entrepreneurship," writes noted marketing expert Peter Drucker, "needs to be based upon purposeful innovation." [2]

The remaking of radio supplies yet another key lesson that can be applied to all service-oriented industries: owners must channel the company's resources to producing or acquiring a superior product (in radio that translates into programming) that will fill a void in a given market. Then consumers will come.

There is also one overriding benefit to this rapidly changing time in radio that should not be overlooked. When future broadcast historians examine this period in radio history, they will see how the new guardians of radio raised the consciousness of the aural medium among both Wall Street financiers and major advertising agencies and advertisers.

> This has always been a fundamentally sound business. It did not mean that cash flows and billing were going up monumentally every year or that there were greater efficiencies every year. It did mean that [in the] longterm this would be a good industry.—Jeff Smulyan[3]

This discussion of radio's late twentieth-century business environment would not be complete without noting the skillful leadership from the industry's long-established broadcast conglomerates. These executives fortified their radio station groups through the medium's sea change during the late 1980s and early 1990s. They include Nancy Widmann, president of CBS Radio Division, who succeeded Bob Hosking in 1989; James Arcara, president of the CapCities/ABC Radio division; and Dick Harris, Jim Thompson, and Dan Mason, all of whom guided Westinghouse Broadcasting's Radio division (Group W Radio) in succeeding order. It was under Harris's watch when the Metropolitan and Legacy stations merged into Group W Radio. (See Chapter 3.)

NEW DIRECTIONS

As radio strengthens, it should again foster program experimentation as it did with the development of all-sports radio. The early nineties have seen new forms of children's radio programming, designed primarily for AM stations, that have been attracting attention among radio operators. Generally speaking, there can be less serious risk-taking in an industry faced with changing economic conditions, prompting executives to follow proven concepts.

Program experimentation is particularly important in a highly fragmented marketplace. Combining formatic elements is one direction taken by WKXW-FM Trenton, NJ. The station has been a ratings success, garnering industry attention with a format that combines issue-oriented telephone talk with oldies music targeted directly at the aging baby-boomer audience. By providing both talk and familiar music, the station has matched programming with listener tastes.

And in search of something distinctive, some modern rock radio stations are going back to the late sixties–style free-form, progressive programming, where music selections are determined by the on-air talent. This approach has been labeled Adult Album Alternative (or Triple A for short) by industry programmers.

Among the more defined radio format categories in growth modes during the nineties are country music (which, according to format tracking sources, emerged as the leading radio format in recent years in terms of number of country-programmed stations), news, talk, urban contemporary, and sports. In particular, there has been a surge in nationally delivered news and business and in Spanish-language programming.

In 1994, Associated Press broadened its radio news service by launching a 24-hour news network as an alternate source for stations unable to afford the high cost of producing their own all-news format. The same year, CBS Hispanic Radio Network, in a joint arrangement with United Press International, expanded its programming base by offering hourly newscasts. (The network also changed its name to CBS Americas.)[4]

Although there is little doubt among radio broadcasters that there is a major shift toward more nationally controlled programming originations, there is general agreement that the industry is large enough to accommodate both local and national programming sources, provided stations do not abandon their mission to serve the local community with local information.

By necessity, we will see many smaller market radio stations combining national programming vehicles such as subscribing to a satellite-delivered format interspersed with local information and play-by-play sports. In this way, they will have tighter control over costs while effectively serving their communities.

> Good stations are more than just jukeboxes. They have to provide other kinds of service such as public affairs programming and have a strong news presence.—Joe Sullivan[5]

New directions in radio also mean new business development programs. For the commercial radio industry to continue its financial growth at a consistent pace it needs to expand both its local and national advertising bases. Station sales managers and account executives must find out as much background as possible about a prospective client's business and marketing needs and show them how radio can fit into their marketing plans. Simply put, it is market-driven selling. The application of new information technology systems will help establish and build upon customer relationships.

From a pure marketing standpoint, the new generation of duopoly radio owners and managers must sell the inherent strengths and marketing capabilities of radio against other local media, such as newspapers, television, direct mail, and cable. In many markets, for example, local newspaper advertising continues to outgross that of local radio. On average, that translates into one or two daily newspapers attracting more available advertising dollars in a given market than several radio stations in the same market. Depending upon market size and the number of media outlets, this entails using one of two basic marketing strategies. Radio executives could either position their station or stations as the primary local medium to buy, or sell them as an essential part of an advertiser's media plan. Both strategies should demonstrate the medium's upside potential between advertising cost and desired consumer audience.

One way for individual radio stations to market themselves to advertisers is to establish a *brand* identity for the format or formats they program, in much the same way as do other service industries, including automotive and packaged goods. Through their own advertising efforts, radio stations should create an identifiable image in the market, differentiating them (in the consumer's mind) from both competing stations and other local media. And with more media competition, there is a growing consensus among radio owners that the industry now needs to sell the quality of its audience more than the quantity.

AT THE CROSSROADS

Radio, television, and cable are all at the crossroads of a telecommunications explosion in the mid-1990s. In a July 1994 interview in *Broadcasting & Cable Magazine*, FCC Chairman Reed Hundt, a former Washington attorney who succeeded Al Sikes as the head of the FCC in President Clinton's administration, said, "The time has come to re-examine, redefine, restate, and renew the social compact between the public and the broadcasting industry." Among Hundt's stated goals for the nineties is to increase minority employment and ownership as well as increase opportunities for women.

THE NEXT CHAPTER: RADIO IN THE TWENTY-FIRST CENTURY

How will local radio broadcasting function as a business in the twenty-first century? There are many visions for a new framework in which radio will operate. Here is one view of how the operational and marketing aspects of radio might change, proposed by *Radio Ink* magazine, that embraces the many facets of a "smart house" concept:

> You'll walk into a building housing twelve radio stations in one location. Though separated by walls, certain employees and functions will be shared. Traffic, engineering, production, dubbing,

copywriting, accounting, cleaning, clerical will be subcontracted to one independent contractor.

One engineering department will oversee the needs of twelve to fifteen stations. High reliability and low failure rate of the equipment will make this possible. In the engineering office a computer will continually monitor all of the studios and all transmitters (which will require only one or two visits annually, again due to high reliability). There will be no tape—everything will be electronic.

Programming will be individually controlled by each company housed in the building; however, a slew of small satellite dishes will adorn the surroundings. There will be a combination of live programming originating from the building and from satellite-delivered programming. Many of the formats will originate via programming services; however, many others will originate from the corporate offices or stations within the corporate chain. Low-cost satellite uplinks will feed each of the stations with one of corporate's successful home-brew formats . . . .

The corporate marketing director will direct (program) localization, research, and promotion efforts. Because of shared expenses, companies will be able to afford top-notch, quality air talent. Additionally, localized news and bits will be fed via the bird (satellite). Localization writers will be assigned the exclusive job of preparing bits and localized news for three or four markets.

Some stations will opt to share talent in the local market and via the bird. They will find that a newsperson can produce a newscast for five stations as easily as one. Air personalities who may or may not be on-site, will not do four-hour shifts in real time. They will be prerecord all breaks in 30 minutes. During the rest of their eight hours, they will be working promotions and attending sales calls.

Stations will be selling more than airtime. They will be selling direct mail, newsletters, magazines, coupon tabloids, and signage. They will be the main supplier of all advertiser needs. All newspaper ads will be designed by an in-house art department . . . .

Salespeople will attend one combined sales meeting weekly, focused on marketing and radio selling. Local sales managers will be responsible for individual station training. However, many stations will be sold by one sales force in combination.

Salespeople will carry laptops. On the spot they can show clients audience shares, qualitative data, and any other research data cross-referenced with station listening estimates. They will be able to access the station's listener database. Orders will be entered on the spot, enabling the station to quote rates, schedule times and share copy ideas . . . .

> All of the single elements explored already exist. The next step will be the combination of all the parts to create the aforementioned scenario. It's smart business. It still allows for competitiveness and will enhance profitability considerably.[6]

Whether or not all the components to this future vision of radio materialize, the fact is that some of these aspects are already occurring. Radio owners are housing new duopoly owned stations under one roof, sharing resources and expenses. In some cases, station facilities are organized by function: programming and production studios might be located on one floor, with sales, marketing, research, and administration on another floor or in another building in close proximity to the studios. Additionally, station sales executives are now conducting presentations with potential advertisers by laptop computer.

In the late 1980s radio fell into an entrepreneurial vacuum, quickly reaching a saturation level in terms of establishing new stations and programming concepts; the overall economy, especially in the media sector, felt the wrath of a major recession. Because of the lessons learned in the eighties, and due to increasing federal deregulation of local and national station ownership limits, the industry is once again poised to become a positive, long-term investment vehicle. Multiple AM and FM same-market ownership permits commercial radio executives to present stronger demographic buys to advertisers, placing radio in a much more formidable position to compete with other media. And in the early nineties, to fund future growth, many of the major station groups that emerged during radio's new entrepreneurial era—including Saga Communications, Infinity Broadcasting, and, in late 1993, Emmis Broadcasting—filed public offerings.

MEDIA POWERHOUSES

As of the mid-1990s, the major network broadcast companies—CBS and NBC in particular—are looking to strengthen their positions within a growing multimedia environment by considering merger and acquisition proposals from other major media conglomerates. The strategy here is to merge complementary media operations, such as a television broadcast company, and a major cable network programmer into a new company, thereby optimizing the business strengths of both parties. For example, in 1994 QVC, which operates a major home shopping cable network, and CBS announced a merger of the two companies. The deal fell apart when Comcast Corporation, one of QVC's key investors opposed the consolidation.

But new corporate powerhouses will be occurring in all aspects of media, including radio. Witness Infinity Broadcasting, which now operates the Westwood One Radio Network. And, at the time when CBS almost merged with QVC, the latter was also looking into a deal to merge its radio division with that of Westinghouse Broadcasting & Cable (Group W Radio). That move would yield a combined Group W–CBS radio company with major station presence in all top 10 markets, especially in the news and talk formats.

ESTABLISHING A BEACHHEAD

Though the radio industry continues to consolidate, there is still the potential for first-time station and group owners, especially for those seeking to establish a portfolio in the smaller or regional markets, to succeed. In such instances, many of the industry's entrepreneurial owners advise, it is critical to learn the business from experienced operators before making the initial purchase. The key is to produce value on a daily basis.

> Radio is a business where you must deliver a product of value to people. If you go out there and say you are going to give the listener what they want . . . you are treating the listener like a valuable customer. Then by delivering those customers to an advertiser, you have created value at the end of the day.—Dick Ferguson, president/CEO of NewCity Communications[7]

ONE FINAL THOUGHT

While many radio station owners are cautiously watching new technologies develop for program transmissions, such as digital audio broadcasting (DAB), the radio industry is now creating its own digital information highway through RBDS and subcarrier technology. Beyond ownership and programming, here exists the beginning of new business and marketing opportunities for radio's next wave of entrepreneurs.

NOTES

1. Based on an interview with Frank Osborn of Osborn Communications, New York.
2. Peter F. Drucker, *Innovation and Entrepreneurship: Practices and Principles*, (Perennial Library, Harper and Row, 1985), 29.
3. Based on an interview with Jeff Smulyan of Emmis Broadcasting, Indianapolis, Ind.
4. "AP Launching 24-Hour News Network," *Broadcasting Magazine*, April 18, 1994, p. 30; "Spanish Radio in U.S. Heats Up the Airwaves," *Radio World*, June 1, 1994, pp 1+.
5. Based on an interview with Joe Sullivan of Joe Sullivan & Associates.
6. Eric Rhoads, "SmartRadio: The Wave of the Future," *Radio Ink*, Nov. 2, 1992, p. 6.
7. Based on an interview with Dick Ferguson of NewCity Communications, Bridgeport, Conn.

Selected Bibliography

The following articles, books, and periodicals are the industry-specific and general business references and source material consulted for the text. The following should serve as a guide for those who wish to further study the management, programming, regulatory, and technical issues that impact radio broadcasting.

The author is a former senior editor of *Radio for Broadcasting & Cable,* the principal reference used in preparing this book.

ARTICLES

Articles are listed by publication and title. Authorship is given when known or noted in the publication.

Advertising Age

"Bates Oks Saatchi Talks," by Stewart Alter, (May 5, 1986), pp. 1+. Related article: "S&S $450 Million Buys Bates," by Stewart Alter, (May 12, 1986) pp. 1+.

"Biggest Global Agency: DDB–Needham–BBDO," (April 28, 1986), pp. 1+.

"JWT Group Sold," (June 29, 1987), pp. 1, 62.

"Ogilvy, WPP Talking Deal," (May 15, 1989), pp. 1+.

"Ogilvy Under WPP Wing," (May 27, 1989), pp. 1+.

Billboard Magazine

"Broadcasters Tune to RBDS for New Revenue Potential," by Phyllis Stark, (June 5, 1993), pp. 1+.
"New Radio Rules Not Set in Stone," (The initial duopoly ownership expansion ruling), by Sean Ross, (April 25, 1992), pp. 6+.

Broadcasting & Cable Magazine

Formerly Broadcasting. (The magazine did not establish full authorship credit lines until 1992.)

"ABC Can Keep Radio Stations," (July 10, 1989), pp. 29–30.
"AM Radio: Survival of the Fittest," (Aug. 14, 1989), pp. 54–57.
"AM Stereo Rears Its Divided Head," by Sean Scully, (April 12, 1993), pp. 61–62. Related article: "FCC Says C-Quam Is AM Stereo," (Nov. 1, 1993), p. 14.
"AP Launching 24-Hour News Network," by Julie Zier, (April 18, 1994), p. 30.
"Beginning a Legacy," (The formation of Legacy Broadcasting), (Nov. 18, 1985), p. 44.
"The Complicated World of Media Brokering," by John Gallagher, (Nov. 11, 1992), pp. 49–50.
"Counting Radio's Assets," (June 4, 1990), pp. 70–71.
"Court Upholds FCC's Striking of Three-Year Rule," (Sept. 3, 1990), pp. 50+.
"CapCities + ABC: A Bold $3.5 Billion Media Marriage Electrifies Industry & Nation," (March 25, 1985), pp. 31–36.
"CBS, Group W Merger Would Make Beautiful Music," (Aug. 1, 1994), p. 14.
"Clear Channel: Sunbelt Success Story," by Peter Viles and Geoffrey Foisie, (July 5, 1993), p. 19.
"DAB: Radio's Uncertain Digital Future," by Sean Scully, (Sept. 23, 1993), p. 46. Related article: "Europeans & the Battle for Worldwide DAB," by Steve Homer, (Sept. 23, 1993), p. 62.
"Expanding Radio's Horizon to Infinity," interview with Mel Karmazin, by Vincent M. Ditingo, (April 11, 1988), pp. 94–97. Related article: "Mel Karmazin and the Infinite Possibilities of Radio," by Peter Viles, (Sept. 6, 1993), pp. 32–43.
"FCC Looks to Raise Caps on Radio Ownership," (April 29, 1991), pp. 19–20.
"Fifty-Nine Percent of Radio Stations Lost Money in '91," by Peter Viles, (July 6, 1993), p. 6.

"The First 60 Years of NBC," (June 9, 1986), pp. 49–64.
"FM: The Great Leaps Forward," (Jan. 22, 1979), pp. 32–39.
"Freer at Last," (The FCC's initial radio deregulation order), (Jan. 19, 1981), pp. 31–34.
"Group W, NBC Radio Marriage Is Off," (Dec. 8, 1986), p. 54.
"Hill Squeezes the FCC's Radio Rules," by Joe Flint and Peter Viles, (Aug. 10, 1992), pp. 4+; Related article: "Radio Rules: Politics Over Principle," by Harry A. Jessell, (Aug. 10, 1992), pp. 5+.
"HLT: A Three-Letter Worry," (June 4, 1990), p. 44. Related article: "Feds Appear Receptive To HLT Guideline Changes," (Oct. 21, 1991), pp. 38–40.
"Hundt's New Deal: FCC Chairman Says Its Time to Revisit Social Compact Between Broadcasting and Public," by Kim McAvoy, (Aug. 1, 1994), pp. 6–8.
"Metromedia Sold in LBO," (March 31, 1985), pp. 33–34. (The Metromedia radio station group was later renamed Metropolitan.)
"Metropolitan Slims Down to Service Debt," (Feb. 29, 1988), p. 63. Related articles: "Sillerman's Radio Reorganization," (July 4, 1988), pp. 39–40; "$300 Million Sale Would Set Radio Only Record," (April 11, 1988), p. 36.
"New FCC Looks at Broadcasting's Oldest Medium," (Nov. 20, 1989), pp. 31–32.
"A New Network Giant: Infinity to Run Westwood & Unistar," by Peter Viles, (Oct. 18, 1993), pp. 40–41.
"Owners Urge FCC to Liberalize Radio Rules," (Aug. 12, 1991,) pp. 48–49.
"Radio Network Seeks Kid-Parent Audience," by Jim Cooper (Aug. 3, 1993), p. 43.
"Radio's Magic Numbers: 30-30," by Joe Flint, (March 16, 1992), pp. 4+.
"Radio Record: $70 Million for WFAN," (Dec. 16, 1991), p. 4; Related article: "Infinity Goes Public: Initial Offering Would Help Finance $70 Million Purchase of WFAN (AM)," (Dec. 16, 1991), p. 18.
"RCA + GE: Marriage in Takeover Heaven"; "NBC Greets GE with Open Arms"; "RCA & NBC: The Way They Were"; "Merger Reaction: Hopes for an Unchanged Course,"(Dec. 16, 1985), pp. 43–48.
"Record Breaking Deal Set in Motion," (April 24, 1989), p. 28. Related article: "Radio's Newest Force: Group W Sets Its Sights on the '90's," (Dec. 11, 1989), pp. 35–38.

"Sheridan, NBN to Combine Radio Companies," (Nov. 4, 1991), p. 50.
"Robert F. X. Sillerman, Fifth Estater Profile," (July 11, 1994), p. 69.
"Sikes Questioned on New Radio Limits: Chairman Backs up Decision to Increase Station Caps and Liberalize Duopoly Restriction During Questioning by Senate Appropriations Committee," by Randy Sukow, (March 30, 1992), pp. 40–41.
"Ted Turner's Bold Bid for CBS"; "What the Rest of the World Thinks of Turner/CBS"; "Ted Turner: Maverick & Media Missionary"; "CBS Begins to Fight Back," (April 22, 1985), pp. 30–35. Related articles: "CBS's $1 Billion Counterattack," (July 8, 1985), pp. 31–32; "Turner Blasts CBS"; "Turner in Trouble?", (July 15, 1985), pp. 30–32; "FCC's Double Whammy to Turner's CBS Hopes," (Aug. 5, 1985), p. 30; "CBS Takes Out an Insurance Policy," (Oct. 21, 1985), p. 36.
"Thomson Homes in on DBS Marketing," by Rich Brown, (Dec. 20, 1993), p. 59.

The BIA Monitor

"Radio Revenues Rebound," (Jan. 17, 1994), p. 1.

BusinessWeek

"Deal Mania: The Restructuring of Corporate America," by Bruce Nussbaum, (Nov. 24, 1986), pp. 74–76.
"Radio's Empire Builders Are Scanning The Dial," by Mark Landler and Mark Lewyn,(June 29, 1992), pp. 58–62.
"The Technology Payoff: Jobs, Jobs, Jobs—Eventually," by Christopher Farrell and Michael J. Mandel, (June 14, 1993), pp. 72–73.
"The Two Real-Life Dramas At CapCities/ABC," by Mark Landler and Ronald Grover, (Aug. 15, 1994), pp. 76–77.

Inside Radio (Daily News Fax Service)

"71% of Stations in Top 10 Markets Are Owned by Groups," (Aug. 29, 1994), p.2.

The New York Times

"Betting on Small Dish TV," by Edward L. Andrews, (Dec. 15, 1993, pp. D-1+.

Radio Business Report

"Duopoly at Mid-Year, 1994," (July 4, 1994), pp. 8–9.
"Miller Kaplan Tests Market Sellout Percentage Reporting System," (Oct. 4, 1993), p. 9.

Radio Ink (formerly *The Pulse of Radio*)

"First Duopoly Wave May Be Cresting," by Reed Bunzel, (April 12, 1993), pp. 11–12.
"Format Frenzy: How Fragmented Is the Future," by Kathy Bachman, (June 7, 1993), pp. 39–42.
"Hands-Free Radio in the High-Tech Age," (May 9–22, 1994), pp. 29–36
"Radio Executive of the Year: Mel Karmazin," interview with the editors, (Jan. 3–16, 1994), pp. 20–26.
"Robert F. X. Sillerman: Radio's Dealmaker", interview with the editors, (Jan. 15, 1990), pp. 22–26. Related article: "An Acquiring Mind: Bob Sillerman, Chairman & CEO of The Sillerman Companies," interview with the editors, (June 7, 1993), pp. 30–34.
"Smart Radio: The Wave of the Future," by Eric Rhoads, (Nov. 2, 1992), p. 6.
"The State of AM," by Kathy Bachman, (April 11–24, 1993), pp. 20–29.

RadioNews (formerly *The Hall Radio Report*)

This publication was acquired by *Radio Business Report* in the mid-1980s.

"FCC Proposes To Eliminate Three-Year Holding Rule," (Jan. 4, 1982), p. 1.
"Gulf United Nears Deal with Infinity Broadcasting to Sell Spin-Off Stations," (Nov. 11, 1980), p. 8.
"WABC-AM New York Plans to Switch to All-Talk in Early May," (March 1, 1982), p. 1.

Radio & Records

"AccuRatings: Challenging Arbitron and the AQH," by Joel Denver, (Oct. 30, 1992), pp. 34–35.

"Determining the Price of Doing Business," by Henry Lawson, (Dec. 31, 1993), p. 14.
"FCC Grants CapCities/ABC Permanent Waiver," (July 27, 1989), pp. 1+.
"Joining the Computer Age," by Cyndee Maxwell, (July 8, 1994), p. 22.
"NAB Endorses In-Band DAB" (Jan. 22, 1993), pp. 1–17.
"Outlook '94: Duopolies, Improved Technology Mark Rockers Wave of the Future," by Cyndee Maxwell, (Dec. 31, 1993), p. 24.
"R & R Survey Probes News on Music Radio in the 90's," (Nov. 27, 1992), pp. 1+.
"Sikes Sees Possible Revision of New Ownership Rules," by Pat Clawson, (April 3, 1992), p. 4.
"Westwood One Restructures into Two Divisions," (April 22, 1994), p. 1+.

Radio World (newspaper)

"DAR Symposium Faces the Future," by Charles Taylor, (April 20, 1994), pp. 1+.
"DCR to Begin Satellite DAR Service," by Frank Beacham, (March 9, 1994), pp. 1+.
"High Quality AM on a Budget," by John Gatski, (March 23, 1993), pp. 1+.
"NRSC to Guide In-Band DAB Testing," by John Gatski, (April 14, 1993), pp. 1+.
"RAB Foresees Wireless As Future Wave," by Randy Sukow, (March 23, 1994), p. 1.
"Spanish Radio in U.S. Heats up the Airwaves," by Alan Haber, (June 1, 1994), pp. 1+.
"WBBR Is First Audio Superstation," by Frank Beacham, (March 23, 1994), pp. 1+.

The Wall Street Journal

"Big Is Back As Corporate America Deals, Buys and Mergers," (Aug. 4, 1994), pp. A1+.
"Triple A Radio Format Attracts Sponsors," by Kevin Goldman, (Sept. 7, 1994), p. B8.

Radio industry events occasionally appear in entertainment business trade magazines such as *Electronic Media* and *Variety*.

BOOKS, PERIODICALS, AND PAPERS

Authorship is given when known. Some materials listed below were used as general references during the writing of this book. Others are the source of specific facts noted in the text.

Abel, John D., Richard V. Ducey, and Mark K. Fratrik, *RadiOutlook: Forces Shaping the Radio Industry*, National Association of Broadcasters, (1988).

Broadcasting & Cable Yearbook, A Brief History of Broadcasting and Cable, R.R. Bowker, (1993).

Champy, James, and Michael Hammer, *Reengineering the Corporation: A Manifesto for Business Revolution*, HarperBusiness, (1993).

Clarke, Arthur C. *How the World Was One: Beyond the Global Village*, Bantam Books, 1992.

De Sonne, Marcia L., *Digital Audio Broadcasting: Status Report and Outlook*, National Association of Broadcasters, (1990).

Ditingo, Vincent M. (with contributions from Pat Healy) "Radio as a Long-Term Investment: A Financial White Paper," The Interep Radio Store, (1991).

The Radio Music Licensing Sourcebook, The Radio Music License Committee and Group W (Westinghouse) Radio, (1990).

Dodd, Mark Reynolds, and Marcia L. De Sonne (ed.), *Satellites and Radio Broadcasting: Historical Review & Market Developments*, National Association of Broadcasters, (1991).

Douglas, George H., *The Early Days of Radio Broadcasting*, McFarland & Company, Inc. Publishers, (1987).

Drucker, Peter F., *Innovation & Entrepreneurship: Practice and Principals*, Perennial Library, Harper & Row Publishers, Inc. (now HarperCollins), (1985).

The Frontiers of Management, 29, (1986).

Duggan, Erwin. S., statement regarding FCC's "Revision of Radio Rules and Policies," (Docket MM 91-141), (Aug. 5, 1992).

The First Fifty Years of Broadcasting: The Running Story of The Fifth Estate (a historical compilation of significant broadcast industry news stories appearing in *Broadcasting Magazine* from 1931 through 1981), Broadcasting Publications Inc., (1982).

Fornatale, Peter and Joshua E. Mills, *Radio in the Television Age*, The Overlook Press, (1980).

A Guide to Understanding and Using Radio Audience Estimates (Arbitron), The Arbitron Company, (1990).

Helregel, Brenda (ed.), *1992 Radio Financial Report*, National Association of Broadcasters, (1992).

Hilliard, Robert L., and Michael C. Keith, *The Broadcast Century: A Biography of American Broadcasting*, (a timeline of historic events in broadcasting), Focal Press, (1992).

Kierstead, Phillip O., *All-News Radio*, TAB Books, (1980).

Kovacic, Rosemary T. Rakas, and Daniel C. Schwartz, "LMAs and the Antitrust Laws," National Association of Broadcasters, (1993).

Krasnow, Erwin G., and Robin B. Martin, *Radio Financing: A Guide for Lenders and Investors*, National Association of Broadcasters, (1990).

Lewis, Tom, *Empire of the Air: The Men Who Made Radio*, Edward Burlingame Books (HarperCollins Publishers), (1991).
"Triumph of the Idol—Rush Limbaugh and a Hot Medium," *Media Studies Journal: Radio: The Forgotten Medium*, Freedom Forum Media Studies Center, (a compilation of writings on radio), Columbia University, (Summer 1993).

Lodish, Leonard M., "Applied Dynamic and Production Models with Specific Application to Broadcast Spot Pricing," *The Journal of Marketing Research*, The American Marketing Association, (May 1980).

MacDonald, Richard J., "On The Business Side—An End to Radio Romance," *Media Studies Journal: Radio: The Forgotten Medium*, Freedom Forum Media Studies Center, Columbia University, (Summer 1993).

McLean, Austin J. (ed.), *RadiOutlook II: New Forces Shaping the Industry*, National Association of Broadcasters, (1991).

Maki, Val, and Jill Pederson, *The Radio Playbook*, Golde Mack, Inc., (1991).

Marino, John. *An Overview of the Radio Broadcast Data System*, National Association of Broadcasters, (March 1, 1993.)

Millman, Nancy, *Emperors of Adland: Inside the Advertising Revolution*, Warner Books, (1988).

Minority Tax Certificate Sourcebook (a compilation of references regarding FCC minority tax certificates prepared in the early 1990s), Minority Enterprise Program, Federal Communications Commission.

Radio Marketing Guide and Fact Book, Radio Advertising Bureau, (1993).

Sanger, Elliott M., *Rebel in Radio: The Story of WQXR*, Hastings House, (1973), 5:46-54.

Sikes, Alfred, FCC statement regarding "Reconsideration of Radio Report & Order," (Docket MM 91-140), (Aug. 5, 1992).

Sklar, Rick, *Rocking America: How the All-Hit Radio Stations Took Over*, St. Martin's Press, (1984).

Webster, Frederick E., Jr., *Market-Driven Management: Using the New Marketing Concept to Create a Customer-Oriented Company*, (part of The Portable MBA Series), John Wiley & Sons, Inc., (1994).

Yates, Joanne, and Robert I. Benjamin, "The Past & Present As a Window on the Future," *The Corporation of the 1990s*, (a compilation of writings on the role of information technology in the corporate workplace), Michael S. Scott Morton (ed.), Oxford University Press, (1991). 3:61-92.

Index

ABC, xx, 1–2, 14, 15
Capital Cities and, 19–20
Rock Radio Network, 13
satellite technology and, 9, 10, 12
talk radio and, 66
ABC Contemporary, 13
ABC Data, 102
ABC Entertainment, 13
ABC–FM, 13
ABC Information, 13
Abel, John, 105
AccuRatings, 32
Acquisitions, 19–22, 77–85, 119–120. *See also specific deals*
local limits, 81
national limits, 81
radio station trading, 84
ripple effect of, 22
Adult Album Alternative, 115
Advertising, 10, 22–27
automating, 97–99
brand identity, 117
commercial load, 34
direct result, 107
expenditures in millions, 47
inflation and, 46
market revenue share, 34–35
national advertising spot clearance time, 11
national market distribution system, 23, 98
run–of–schedule, 33
seasonality, 39
Advertising agencies, xviii, 22–27
foreign controlled, 26
group–owned representation, 24
mega–reps, 24–27
Affiliates, 10, 70
local marketing agreements, 70, 85–87, 106
multiple, 15
special–program, 14
Allen, Fred, xviii
Allen, Gracie, xviii
Alwadish, David, 107
AM–FM Buyer's Worksheet, 98
AM/FM home receivers, xxi
AM radio, 59–73
continuity in, 69
formats of, xxiii. *See also* Radio programming
future outlook of, 72–73
licensing issues of, xxii
noise interference and, 72
ratings and, xii. *See also* Ratings
standard wave of, xviii, 72
in stereo, 70–71

AM radio (*Continued*)
talk radio, 19
AM stereo sets, 71
AMAX, 72
American Marconi Telegraphy Company, xvi
American Society of Composers, Authors, and Publishers. *See* ASCAP
American Urban Radio Network, 15
Amway Corporation, 14
Anti–trafficking rule, xxii, 3
Apollo Radio, 2
Arbitron, 31–32
Arcana, James, 114
Armstrong, Edwin H., xix, xviii, 18
ASCAP, xvi, xvii, xx
Associated Press, 115
satellite technology and, 9
Audience, 3. *See also* Ratings
out of the home, 33
share by age, 33
Audion, xvi
Automated tape and cart machines, 6
Automation, 97–99. *See also* Information technology

Backer, Spielvogel & Bates, 25
Bank of New England, 47
Banks, 40–41, 51–52, 78
Banner Radio, 25
BBDO International, 25
BCFM, 35, 62, 78
Beaudin, Ralph, 13
Belar, 70
Benjamin, Robert I., 97, 99
Benny, Jack, xviii, 5
Black Monday stock market crash, 51
Blair Radio, 25
Blaupunkt, 108
Block, Martin, xix
Bloomberg, Michael, 110
Bloomberg Business Radio, 110
BMI, xix, xx
Brand identity, 117
Brazell, Carl, 48, 50
Breakfast Club, 92
Brenner, David, 65
BROADCAP, 42
Broadcast–Cable Financial Management Association. *See* BCFM
Broadcast cash flow, 39
Broadcast Investment Analysts, 69, 84
Broadcast licensing, xvii
Broadcast Music Inc. *See* BMI
Broadcasting
direct (to home) broadcasting, 110–111
election returns, xvi
Hindenburg dirigible disaster, xix
live broadcasting, xvi
live music, xvii
news, xx
"War of the Worlds," xix
Broadcasting & Cable, 52, 54, 117
Burns, George, xviii
Business Radio Network, 12
BusinessWeek, 99

Cable audio, 110–111
Cable News Network. *See* CNN
Capital Cities, 19
Caruso, Enrico, xvi
Cash flow, 53, 79, 113
CBS, xviii, xx, 2, 15, 119–120
all–new format and, 64
RadioRadio, 13
satellite technology and, 9, 10, 12
sports programming and, 19

Ted Turner and, 21
Tisch and, 21
CBS Americas, 115
CBS Hispanic Radio Network, 115
CBS Radio Representative, 26–27
"Chain Broadcasting," xx
Chain rule, xvii, xx, xxii
Chicago, Illinois
all–new formating in, 64
Christian, Ed, 49, 56, 82
CIGNA, 46, 47
Clark, Dick, 12, 14
Clear Channel Communications, 57
Clinton, Bill, 17
Closed–circuit music, xxi
CNN, 16, 21
Columbia Broadcasting System. *See* CBS
Columbia Phonograph Broadcasting System, xviii
Combos, 62
Comcast Corporation, 120
Command Communications, 50
Commercial load, 34
Communications Act of 1934, xviii
Competitive Media Reporting, 100–102
Consolidation, xi, 14, 82, 114–115, 119–120
Consumer Price Index, 46
Contemporary hit radio, 12
CouponRadio, 107
Crosby, Bing, xviii
Cross–ownership rule, xxiii, 20

D & R Radio, 25
Daren F. McGavren Company, 24
DATS, 10
De Forest, Lee, xvi
Delco Electronics Corp., 108
Demographics, 60, 63, 89–90
Denon, 107, 108
Dentsu, 26
Department of Commerce, xvii
Deregulation, xi, xii, 2–28
AM stereo and, 71
annual sales and, 46
public affairs programming and, 15
Digital audio broadcasting, xiii, 108–110
Digital Audio Transmissions System. *See* DATS
Digital broadcast technology, 105–110
Digital cable audio, 110–111
Digital phone, 111
Dingell, John, 80
Diode tube, xv
Direct broadcast satellite, 110–111
Direct (to home) broadcasting, 110–111
DirecTV, 110, 111
Disc jockey, xix
Docket 56, 61, 80–90
Donovan Data Systems, 98
Doubleday, 49
exit from radio, 4
Doyle Dane Bernbach, 25
Drexel Burnham Lambert, 51
Drucker, Peter, 114
Duggan, Erwin, 78, 80
Duncan's American Radio, 35
Duopoly, 62, 78–85, 116
personnel and, 87–89
Durpetti & Associates, 25

Easy listening, xxiii
Economies of scale, xiii
Edens, Gary, 49
Edens Broadcasting, 2, 49
EFM Media, 67
Electromagnetic waves, xv
Electronic mail, 102
Emmis Broadcasting, 2, 51, 67, 120
acquistions and, 22

Entrepreneurs, xii, 2, 45–58, 113–115
Equity, 41
Evergreen Media, 2

FCC, xviii
AM radio and, xxii
anti–trafficking rule, xxii, 3
"Chain Broadcasting" report of, xx
chain rule and, xvii, xx, xxii
cross–ownership rule, xxiii, 20
deregulation and, 2–28
Docket, 56, 61, 80–90
duopoly rule and, 78–85
FM radio and, xx, xxi–xxii
one–to–a–market rule, xxii–xxiii, 20
regulation of ownership, xx
30/30 rule, 79
three–year rule, 5
Federal Communications Commission. *See* FCC
Federal Radio Commission. *See* FRC
Federal Reserve Board, 52
Ferguson, Dick, 48, 120
Fessenden, Reginald, xvi
Financial issues, 2, 31–58, 113–121
annual sales, 46
broadcast cash flow, 39
budgets, 35–38
cash flow, 53, 79, 113
lenders, 40–41, 51–52, 78
market revenue share, 34–35
profit, 38–39, 62, 78
resale values, 39–40
salaries, 36–38
tax certificates, 41–42
Fleming, John A., xv
FM radio, xix, xviii, xxi
adult rock format, 18
album–oriented rock, 18
versus AM, 59–73
channel assignments and, xii
digital broadcasting and, 105–106
FCC and, xx, xxi–xxii. *See also* FCC
formats of, xxiii. *See also* Radio programming
frequency of, xx
IBOC systems and, 109
popularity of, 18–19
progressive rock and, xxii, 18
stereo broadcasting, xxi, 18
FM radio receivers, xix
Foote, Cone & Belding, 26
FRC, xvii
Freed, Alan, xxi
Fritts, Eddie, xi

GE, xvi
acquistions and, 22
FM radio receivers and, xix
as owner of NBC, xvii
RCA and, 20–21
General Electric. *See* GE
General Electric SATCOM Satellite System, 7
General utilities rule, 49
Gerry House & The House Foundation, 94
GM Hughes, 110
Goldstein, Steve, 77
"Great Raid of '48, The," xx
Gross, Gil, 67
Group W Radio, 50, 114, 115, 120

Harris, Dick, 114
Harris Corp., 70
Harte–Hanks Communications
exit from radio, 4
Harte–Hanks Radio Group, 49

Hartford, 46
Harvey, Paul, 19
Hertz, Heinrich Rudolf, xv
Highly leveraged transactions. *See* HLTs
Hiller, Newmark, Wechsler & Howard Agency, 24–25
Hindenburg dirigible disaster, xix
Hirsch, Carl, 49
HLTs, 51–52
Hoberman, Ben, 66
Hoffman/Schultz Media Capital, 62
Holding company, 26
Hollings, Ernest, 80
Home Shopping Network, 90
Hope, Bob, xviii, 5
Hosking, Bob, 114
Hubbard Broadcasting, 110
Hundt, Reed, 117

IBOC, 108–109
Imus, Don, 16, 17, 93
Infinity Broadcasting, 2, 14, 16–17, 56–57, 67, 88, 120
Information technology, xiii, 97–112
  local stations and, 99–102
Integrated Services Digital Network, 111–112
Interep Radio Store, 24, 26–27, 38, 98
Interpublic Group of Companies, 26
Investments, 48–53, 79, 113–114. *See also* Financial issues

J. Walter Thompson, 26
Jarvis, Al, xix
Joe Sullivan & Associates, 87
John Blair & Co., 4
Joint operating agreements, 86
Jones Satellite Services, 12
Josephson International, 49
  exit from radio, 4
Journal, 52
Junk bonds, 51

KABC–AM Los Angeles, 65, 66
Kahn Communications, 70
Karmazin, Mel, 14, 56–57, 88
Kasem, Casey, 92
Katz Broadcasting, 48–49
Katz Communications, 25
  exit from radio, 4
Katz Radio Group, 24, 25, 26–27
Kaufman, Murray, 92
KDKA–AM Pittsburgh, xvi, xvii, 62, 69
KFWB Los Angeles, xix
King, Larry, 65
KJOI–FM Los Angeles, 50
Kluge, John, 48
KMOX–AM St. Louis, 69
KSLX–FM Los Angeles, 93
KYW–AM New York, 64

Laptop computers, 119
Larry King Show, 65
Lawson, Henry, 38
Legacy Broadcasting, 2, 50
Lending retreat, 51–52
Lewis, Tom, 67
Liddy, G. Gordon, 67, 95
Limbaugh, Rush, 67, 73, 95
LIN Broadcasting, 4
Lintas Worldwide, 26–27
Live broadcasting, xvi
Local area networks, 99
Local marketing agreements, 70, 85–87, 106
Locally-produced programming, 12
  versus national, 16
  radio news staff, 15

MacDonald, Richard J., 52
Magee, William, 49
Magnavox, 70, 71
"Make Believe Ballroom," xix
Malrite Communications, 49
Marconi, Guglielmo, xv
Marconi Telegraphy Company, xv
Market revenue share, 34–35
Market Sellout Percentage, 100
Marketing issues, 2, 118–119. *See also* Advertising
Mason, Dan, 114
McCann–Erickson agency, 26–27
McGavren Guild Radio, 24
McLaughlin, Edward F., 67
McLendon, Gordon, xxi, 63–64
McNeill, Don, 91
Media conglomerates, 1
Mercury Theatre of the Air, xix
Mergers. *See* Acquisitions; Consolidation; *specific deals*
Metromedia, 2, 4, 48, 49, 50
Metropolitan Broadcasting, 48, 50
Metropolitan Opera House, xvi
Miller, Kaplan, Arase & Co., 35, 100
Minority ownership, 41–42
Minority Telecommunications Development Program, 42
Moby in the Morning, 93
Monopoly issues. *See* Consolidation; FCC
Morgan Stanley & Co., 46
Morgan Stanley and CIGNA Leverage Capital Fund, 47
Morning radio syndication, 91–95
Morrison, Herb, xix
Morrow, Bruce, 49
Morse Code, xv
Motorola, 70–71
  C–Quam system, 71
Multiple "chain" station, xvii
Multiple network concept, 13
Music Choice, 111
Music license fees, xix, xvii, 36
Music licensing organization, xvi, xx
Mutual Broadcasting, xviii, 12, 14, 65
  satellite technology and, 9

NAB, xi, xvii, xxi, 35, 36, 50, 62, 78, 80, 87, 106
NAFMB, xxi
National Association of Broadcasters. *See* NAB
National Association of FM Broadcasters. *See* NAFMB
National Black Network, 15
National Broadcasting Company. *See* NBC
National News and Information Service, 11
National programming
  versus local, 16–17, 116
National Public Radio
  satellite technology and, 8–9
National Radio Broadcasters Association. *See* NRBA
National Radio Standards Committee. *See* NRSC
NBC, xvii, 2, 14, 21, 119–120
  acquistions and, 22, 119–120
  exit from radio, 4
  as monopoly, xx
  satellite technology and, 9, 10, 12
  Source, The, 13
  talk radio and, 66
  twenty–four hour format network, 11
NBC Blue Network, xvii
NBC Red Network, xvii
Needham Harper Worldwide, 25
Networks, 10–17
  affiliates and, 10

demographics and, 15
formats and, 11–13
New York, New York
all–new formating in, 64
blackout coverage, 64
New York Times, 109
NewCity Communications, 2, 49
Noble, Edward J., xx
Noble Broadcast Group, 2
Nouveau radio, 45–48
Nouveau radio groups, 5
NRBA, xxi
NRSC, 72, 108

Ogilvy & Mather, 26
OmniAmerica Communications, 50
Omnicom, 25
One–to–a–market rule, xxii–xxiii, 20
Optic transmissions, 111–112
Osborn, Frank, 114
Osborn Communications, 2
Osgood, Charles, 19
Owens, Gary, 92
Ownership. *See* Acquisitions

Paley, William, xviii, 21
Pattiz, Norm, 13
Paul Kagan Associates, 3, 83
Paxson, Bud, 90
Personnel
duopolies and, 87–89
salaries, 36–38
Pollack Media Group, 94
Power ratio, 34–35
Premier Radio Networks, 93
Program carriage fees, 11
Project Acorn, 109
Public Broadcasting Service, 80

QVC, 120

Racco, Al, 66
RADAR audience survey reports, 33, 60
Radio. *See also specific topics*
nouveau radio, 45–48
timeline of, xiii–xxiv
Radio Act of 1927, xvii
Radio Advertising Bureau, 2
Radio broadcast data systems. *See* RBDS
Radio Corporation of America. *See* RCA
Radio frequencies, xv. *See also* AM radio; FM radio
Radio industry. *See also* Financial issues; Investment
compared to real estate, 54
competition among stations, 3
entrepreneurial consolidation, 4
future outlook of, 72–73, 115–121
investment value of, 3. *See also* Investment
leaders in, 114
revenues, xxi, 31–35, 38–39, 55
Radio Ink, 88, 117
"Radio music box," xvi
Radio programming, 118
Adult Album Alternative, 115
big band, 69, 90
block programming, 5
business, 63, 115
contemporary hit radio, 12
country music, 115
demographics and, 60, 63, 89–90
disco, 66
duplication of FM and AM, xxii, 61, 84
fragmentation, 63
informational programming, 63
morning programming, 16, 18
motivational, 63
national vs. local, 16–17, 116

Radio programming (*Continued*)
news, 63–64, 115
progressive rock, xxii, 18
rock 'n roll, xxi, xxii, 6, 59
Spanish, 115
specialized, xxi, 5–6, 7, 63–65
sports, 63, 67, 68, 115
talk radio, 19, 65–69, 115
top 40, xxi, xxii, 6, 68, 92
twenty–four hour format network, 11
urban contemporary, 115
variety of, 61, 81
vertical vs. horizontal, 89–90
weather, 63
Radio receivers, xix, xxi, 108, 109
Radio stations. *See also specific stations*
government classification of, xvii
local, 99–102
resale values, 39–40
trading of, 84
Radiocard, 107
RadioRadio, 13
Raphael, Sally Jesse, 66
Ratings, 3, 31–34, 94
trends in AM and FM, 60
RBDS, 106–108
radio receivers, 108
RCA, xvi
GE and, 20–21
as owner of NBC, xvii
Reagan, Ronald, 2
Recording coverage, xix
Remote coverage, xix
Risk capital, 41
RKO Radio Network, 12, 14
satellite technology and, 9
Robinson, C. T., 11
Rock Radio Network, 13
Run–of–schedule, 33
Saatchi, Charles, 25
Saatchi, Maurice, 25
Saatchi & Saatchi P.L.C., 25
Sabo, Walter, 91
Saga Communications, 2, 49, 56, 119
Salaries, 36–38
Sales power, 98. *See also* Advertising
Salespeople, 118–119
laptop computers, 119
Sarnoff, David, xix, xvi
SATCOM, 10
Satellite Music Network, 2, 11, 14, 51
Satellite technology, xi, xii, 2, 6–10
Schwartz, Walter, 13
Scientific Atlanta, 10
SFX Broadcasting, 50
Shamrock Broadcasting, 49, 88
Sheridan Broadcasting Network, 15
Sikes, Alfred, 82, 117
Sillerman, Robert F. X., 49
Sillerman–Magee Communications Management, 49, 50
Smart house concept, 117–118
Smartcard, 107
Smith, Kate, xviii
Smulyan, Jeff, 51, 79
Snyder, Tom, 66
Sorrell, Martin, 26
Source, The, 13
Spanish Broadcasting System, 2
Spanish–language programming, 115
Specialization. *See also* Radio programming
satellite technology and, 9
Spot advertising, 23
Stakelin, Bill, 2
Stand alone properties, 62

Statistical Research, Inc., 33, 60
Stern, Howard, 16, 17, 93, 94
Stevens, Gary, 54
Storz, Todd, xxi
Strategic Radio Research, 32
Subsidiary Communications Authorization, xxi
Sullivan, Joe, 116
Sutter, Diane, 88
SyndiNets, 17, 91–95

TA Associates, 47
Talk radio, 65–69
Tape/cartridge systems, xxiii, 11, 12
Tax certificates, 41–42
Tax codes, 49
Ted Bates Worldwide, 25
Telephone lines, xvii, 8, 111
Television, 10
  programming of, xxi
  revenues of, 1
Texas State Networks, 50
Thomas, Lowell, xvii
Thompson, Jim, 114
Thompson Consumer Electronics, 111
Three–year rule, 5
Tisch, Lawrence, 21
Tom Joyner Show, 93
Trans–Atlantic wireless signal, xv
Transistor, 6
Transtar Radio Network, 11, 14
Turner, Ted, 21
Turner Broadcasting System, 21
Twenty–four hour format network, 11
Tyler, John, 11
Tytel, Howard, 49

Unistar, 2, 14, 93
United Press International Audio Networks
  satellite technology and, 9
United States Satellite Broadcasting, 110
United Stations, 12, 14
United Stations syndication company, 14

Vacuum tube, xvi
Verbitsky, Nick, 12, 14

WABC–AM New York, xxii, 64, 65–66, 67
"War of the Worlds," xix
WASH–FM Washington, D.C., 50
WBBR–AM New York, 110
WBZ–AM Boston, 69
WBZ Springfield, MA, xvii
WCAU–AM Philadelphia, 69
WCCO–AM Minneapolis–St. Paul, 69
WEAF New York, xvii
Welles, Orson, xix
Wells, Rich, Greene, 26
Welsh, Pat, 94
WESTAR satellite
  satellite technology and, 9
Westinghouse, xvi, 2, 50, 64, 114
  acquistions and, 22, 120
  as owner of NBC, xvii
Westwood One, 2, 13–14, 14, 15, 51, 120
  acquistions and, 22
WFAN–AM New York, 17, 51, 67, 68, 73, 93
WGN–AM Chicago, xviii, 69
WHN–AM New York, 67
Widmann, Nancy, 114
WINS–AM New York, 64

WIOD–AM Miami, 65
WIP–AM Philadelphia, 50
Wire Products & Baskets, 26
Wireless receiving sets, xvi
WJFK–FM Washington D.C., 67, 93
WJIT–AM New York, 57
WJR–AM Detroit, 69
WKHX–FM Atlanta, 93
WKTU–AM New York, 57
WKXW–FM Trenton, NJ, 115
WLS Chicago, xix
WLW–AM Cincinnati, xviii, 69
WMAL–AM Washington, 69
WMCA–AM New York, 64
WNAC Boston, xvii
WNBC–AM New York, 68
WNEW New York, xix
WOR–AM New York, 69
WOR Newark, NJ, xviii
WQXR New York, xix
WSIX–FM Nashville, 94
WTBS–TV Atlanta, 21
W2XMN–FM Alpine, NJ, xix
WXRK–AM New York, 17, 57
WXRK–FM New York, 93
WXYZ Detroit, xviii
WYSP–FM Philadelphia, 57, 93

XETRA–AM Tiajuana, Mexico, xxi, 63

Yates, Joanne, 97, 99
Yield management, 99–100

Zenith system, 71